Foraging Theory

MONOGRAPHS IN BEHAVIOR
AND ECOLOGY

Edited by John R. Krebs and
Tim Clutton-Brock

Foraging Theory

DAVID W. STEPHENS
and JOHN R. KREBS

Princeton University Press
Princeton, New Jersey

Contents

Preface

> When we contemplate every complex structure and instinct as
> the summing up of many contrivances each useful to the pos-
> sessor, nearly in the same way as when we look at any great
> invention . . . how far more interesting will the study of natural
> history become!—Darwin, *On the Origin of Species*

This book analyzes feeding behavior in the way an engineer might study
a new piece of machinery. An engineer might ask, among other questions,
about the machine's purpose: is it for measuring time, wind speed, or income tax? This is a worthwhile question for the engineer because machines
are built with a purpose in mind, and any description of a machine should
refer to its purpose. Asking what the machine is for helps the engineer
understand how it works. To give a trivial example, we would find it easier
to work out how a slide rule operates if we knew it was meant for doing
calculations and not for digging holes.

Biologists also ask questions about purpose, about what things are
for. For example, Lewontin (1984) says: "It is no accident that fish have
fins, aquatic mammals have altered their appendages to form finlike flip-
pers, . . . and even seasnakes, lacking fins, are flattened in cross-section.
It is obvious that these traits are adaptations *for* aquatic locomotion"
(emphasis ours). In contrast to the engineer, the biologist thinks of design
or purpose as the product of natural selection, rather than as the product
of a conscious creator. Natural selection chooses traits that are useful in
the struggle for survival and reproduction. A lion seems well designed for
killing gazelles because traits that make lions good gazelle-killers were
useful to the lion's ancestors: they allowed the lion's ancestors to produce
more offspring than were produced by lions with other traits.

Design or adaptation is related to fitness (survival and reproductive
success) but analyzing design is not the same as measuring fitness. If one
attempted to study adaptation simply by measuring survival and reproductive success, one would reach the vacuous conclusion that those that
survive and reproduce are those that survive and reproduce (Scriven 1959,
Beatty 1980). Even showing that fitness varies between individuals with
different traits is not enough to infer adaptation; one must know how the
traits influence fitness. In other words, the central question in the study
of adaptation is not just whether individuals survive, but how *design* is
related to expected survival and reproduction (Mills and Beatty 1979).

Williams (1966) makes this point when he says that measuring reproductive success

focuses attention upon the rather trivial problem of the degree to which an organism actually achieves reproductive survival. The central biological problem is not survival as such, but design for survival. (p. 159)

The study of adaptation, therefore, is an integral part of evolutionary biology, and models of design such as optimality models are not merely shortcut versions of genetic models, as Lewontin (1979) has suggested, nor are they simply a staging post toward the measurement of fitness. They are part of a separate and necessary enterprise: biologists must study the usefulness of traits if the theory of natural selection is to explain adaptation (Beatty 1980).

Optimality modeling, the theme of this book, is one method that raises the study of design from clever "story telling" (Gould and Lewontin 1979) to a position in which "explicit, quantitative and uncompromising" hypotheses allow biologists to "recognize logical implications or to demand that there be a precise congruence between theory and observation" (Williams 1966, p. 273). Other approaches include comparisons between or within species (e.g. Ridley 1984) or between different experimental treatments (Tinbergen et al. 1967). The arguments in favor of and against optimality models are discussed in Chapters 1 and 10 respectively, and we will not repeat them here. Instead, we will briefly sketch a few discoveries that have, in our view, come directly from optimal foraging theory.

First, some examples of phenomena that were already well known, but whose significance was obscured because they lacked a cogent theoretical interpretation. Psychologists knew that animals were sensitive to variance as well as mean reward for at least 20 years before Caraco et al. (1980b) used a foraging model to explain it (Chapter 6). This explanation provided a way of organizing existing evidence, and it made bold and unexpected predictions about which factors should influence animal sensitivity to variance (see below). "Wasteful killing" or "partial consumption of prey" was also well known by students of behavior before foraging theory came to light. But now, instead of viewing it as an oddity or maladaptive peccadillo, behavioral ecologists can make use of economic considerations to account for its occurrence, and they can successfully predict just how wasteful the forager should be (e.g. Cook & Cockrell 1978).

What about phenomena that foraging theory has revealed or highlighted? It is of course impossible to claim that a single approach was the stimulus for any particular discovery, but there seems little doubt that foraging theory played a major role in enabling biologists to discover how pollinator foraging behavior affects the design of plants (Pyke 1978a, Best and Bierzychudek 1982), and that what are essentially foraging models were important in the discovery of individual variation in mating strategies (Parker 1978).

Furthermore, foraging models have predicted effects and phenomena whose occurrence was not predicted by other theories. The prediction that an animal's energy budget influences its sensitivity to variance in reward is one striking example (Chapter 6), as is the prediction of environmental conditions under which a foraging animal should and should not show exploratory or sampling behavior (Chapter 4). More specifically, optimal foraging models can generate predictions that run counter to currently accepted psychological theory (and that are borne out by observation) about the conditions under which an animal's choice should not minimize the delay until the next reinforcement (Houston 1986). These examples show that foraging theory is more than an elegant technique, that it has provided and will continue to provide insight into why animals behave the way they do.

Lastly, a few words about the contents of this book. Chapter 1 explains the basic rationale of optimality models. We follow this explanation with three chapters on models that maximize the long-term rate of energy intake. Chapter 2 deals with the classical prey and patch models, and Chapter 3 examines modifications of these models: what happens if the forager encounters more than one prey item at a time, for example. In Chapter 4 we consider information, viewing learning from the perspective of gaining and using information economically. This approach differs greatly from the way psychologists usually analyze learning.

In Chapter 5 we examine currencies other than energy gain; economic models of complementary resources offer a potential way to analyze "mixed currencies," but they have seldom been applied in behavioral ecology. Behavioral ecologists might use these models to analyze herbivore diets, although we conclude that for this purpose simpler modeling approaches may be adequate. Chapter 6 discusses the major alternative currency to rate-maximizing in foraging models: minimizing the likelihood of energetic shortfall. Chapter 7 briefly discusses dynamic optimization as a way to model complex extensions of foraging such as daily time budgets and life history tactics. The final three chapters are not directly concerned with theory. In Chapter 8 we introduce the idea that animals may use "rules of thumb" to solve foraging problems. Chapter 9 presents a detailed review of the evidence for and against the basic prey and patch models. The available evidence teaches the empiricist a salutary lesson: one should make sure that the assumptions of the model being tested are actually met. We return to generalities in Chapter 10, in which we try to answer some of the criticisms of the optimality approach. We conclude that one must compare observed design to an optimality model to find out whether phylogeny, genetics, and ontogeny constrain the design of organisms. So, these criticisms, rather than serving as arguments against the optimality approach, highlight one of its uses.

Our coverage of topics is necessarily uneven. For some topics we attempt a synthesis, for others we simply review the literature. Although we believe this unevenness reflects the state of the art as much as it reflects our own biases, we hope that the reader will be stimulated to develop those parts of the subject that we analyze superficially.

Acknowledgments

This book is distinguished, if by nothing else, by its long and geographically broad gestation and birth. It was conceived in Oxford in late 1982. The detailed outline and initial writing were accomplished in spring 1983, during a visit by JRK to the Smithsonian Environmental Research Center, where DWS was a Smithsonian Visiting Scientist. The embryonic manuscript was then carried north of the border to Vancouver, where DWS held a NATO postdoctoral position at the University of British Columbia. At the same time, JRK was an International Scientific Exchange Visitor at the University of Toronto, and a considerable part of the text was written during a visit by DWS to Toronto in spring 1984. By now the manuscript was substantial, but it had to make two more trips before the first draft was complete. DWS took it to Salt Lake City, where he is an NSF postdoctoral fellow, and the finishing touches were added during a short session at the University of California at Davis, where JRK was visiting as Storer Lecturer. We are grateful to all these institutions that have wittingly or unwittingly acted as our hosts.

We are also grateful to the following people who helped and tolerated us during our fits of writing: Jim Lynch and Gene Morton of the Smithsonian Institution; Lee Gass of the University of British Columbia; Ron Ydenberg and Larry Dill of Simon Fraser University; Jerry Hogan, Sara Shettleworth, and David Sherry of the University of Toronto; Ric Charnov of the University of Utah; and Judy Stamps of the University of California at Davis.

Many people contributed to the manuscript by commenting on drafts, sharing ideas, or showing us unpublished work. We are especially grateful to Dawn Bazely, Tom Caraco, Ric Charnov, Innes Cuthill, Tom Getty, Jim Gilliam, Alasdair Houston, Alex Kacelnik, Steve Lima, Jim McNair, Naomi Pierce, Sara Shettleworth, Anne Sorensen, and Ron Ydenberg. We thank Clare Matterson for drawing the figures.

We acknowledge a special debt to Ric Charnov and Gordon Orians. Ric Charnov introduced us both at different times to foraging theory. Gordon Orians's encouragement helped to convince us that this project was worthwhile. Charnov and Orians's unpublished book on foraging theory provided the foundation for our efforts.

DWS thanks Kathy Krebs for her kindness during his visit to Toronto. DWS also thanks Anne Sorensen for putting up with three years of book writing.

In addition, we thank the following publishers for permission to reproduce figures from their publications: Bailliere Tindall for permission to reproduce Figures 1.1, 6.5, B6.2, 7.1, and 9.1, all of which originally appeared in *Animal Behaviour*; University of Chicago Press for permission to reproduce Figures 3.7 and B5.1, which appeared in *The American Naturalist*; and Cambridge University Press for permission to reproduce Figures 5.4 and 5.5, which appeared in *Behavioral and Brain Sciences*, and Figure 8.2, which appeared in *Adaptive Behavior and Learning* by J. E. R. Staddon.

We also thank the following authors for permission to reproduce figures from their work: J. B. Wallace for permission to reproduce Figure 1.1, which originally appeared in Wallace and Sherberger (1975); E. L. Charnov for permission to reproduce Figure 3.5, which originally appeared in Charnov and Orians (1973); T. Getty for permission to reproduce Figure 3.7, which originally appeared in Getty (1985); H. Rachlin for permission to reproduce Figures 5.4 and 5.5 from figures which originally appeared in Rachlin et al. (1981); B. Winterhalder for permission to reproduce Figure B5.1 from figures which originally appeared in Winterhalder (1983); T. Caraco for permission to reproduce Figure 6.5 from figures which originally appeared in Caraco et al. (1980b); J. H. Kagel for permission to reproduce Figure B6.2 from figures which originally appeared in Kagel et al. (1986); R. C. Ydenberg for permission to reproduce Figure 7.1 from figures which originally appeared in Ydenberg and Houston (1986); J. F. Gilliam for permission to reproduce Figures 7.2, 7.3, and 7.4 from figures which originally appeared in Gilliam (1982); J. E. R. Staddon for permission to reproduce Figure 8.2 from a figure which originally appeared in Staddon (1983); and A. Kacelnik for permission to reproduce Figure 9.1 from a figure which originally appeared in Kacelnik and Houston (1984).

Salt Lake City, USA
Oxford, England
August 1985

Foraging Theory

Foraging Economics: The Logic of Formal Modeling

1.1 Introduction

Some caddisfly larvae spin silken catch-nets. These nets capture small plants, animals, and organic particles that are swept into them by the streams in which the larvae live. The larvae's nets (and often their bodies) are fixed to some immobile object in the stream such as a rock or submerged tree trunk. The nets are not sticky or electrostatically charged: they simply stop particles that are too big to pass through the mesh (Georgian and Wallace 1981). The net-spinning caddisflies have capitalized on their moving medium in an elegant and apparently straightforward way. They have built foraging sieves.

Caddisfly nets are astonishingly diverse. They vary in size, shape, and location in the stream, and, spectacularly, in the structures built to support the net. Wallace and Sherberger (1975) have appropriately described the net and accompanying structure of *Macronema transversum* larvae (Fig. 1.1) as "possibly one of the most complicated feeding structures constructed by non-social insects." Students of caddisflies (see Wallace and Sherberger 1975) believe that this structure takes advantage of subtle hydrodynamic principles (the law of continuity will slow down the flow across the net in comparison with the flow in the entrance and exit ports; the Bernoulli effect—of water moving across the exit port—will drive water through the structure). Among the simpler net designs are the large-meshed, round, trampoline-shaped nets of most hydropsychid larvae and the long windsock-shaped nets with fine meshes built by philopotamids.

Non-adaptive variation might explain the variety and detail of caddisfly nets, but consider for the moment that net structure reflects the action of natural selection. How then can we interpret the element of its design, for example, its mesh size and shape?

A first step is to find out whether there are systematic trends linking mesh size and shape with environmental factors. A comparative survey shows two trends. First, larger meshes are associated with faster water. Second, size is correlated with shape. Small meshes are usually long and rectangular, and larger ones are roughly square. Both trends can be interpreted in terms of costs and benefits (Wallace et al. 1977). One hypothesis

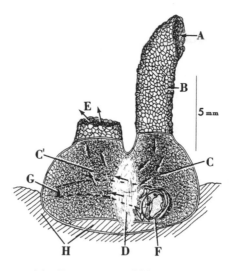

Figure 1.1 Catch-net and feeding structure of *Macronema transversum*. (A) Entrance hole (facing upstream) for in-flowing water. (B) Sand and silk entrance tube. (C) and (C') Anterior and posterior portions of chamber. (D) Capture net. (E) Exit hole for out-flowing water. (F) Anterior opening of larval retreat with larva in place. (G) Exit hole from larval retreat (for faeces and water flowing over gills). (H) Substrate. Arrows represent direction of water flow. Broken lines between (F) and (G) signify approximate position of larval retreat.

is that large meshes can withstand swifter currents because they present less resistance to the flow. A second possibility is that smaller nets are better in slow water because slow streams carry smaller particles. Furthermore, in slow streams fewer liters of water per minute pass through the nets, so the larvae may need to capture a larger proportion of particles to meet their food requirements. Another hypothesis is that the size and shape of the mesh are correlated because of the cost of silk. Caddisfly larvae must use a much greater length of silk to fill a given area with meshes if the meshes are small, but the extra cost can be reduced by making the meshes long and narrow, cutting the cost of cross pieces. This saving may not be possible with larger meshes because the silk stretches when a large particle collides with the net, turning the rectangular mesh into a distorted and ineffective hexagon.

These ideas sound reasonable, but they might be criticized for being no more than plausible stories. They certainly do not meet Williams's (1966) criterion of an "explicit, quantitative, uncompromising" design hypothesis. Formal models may help biologists to evaluate design hypotheses by helping to analyze the problem and by making testable predictions.

To analyze a design problem is to break it into parts and to determine the relationship among the parts. A formal analysis clarifies what the parts of the problem are, and it reveals their full implications and interactions. What relationship between water velocity and mesh size would be expected from the "resistance to breakage" hypothesis? Could the argument about the silk-stretching effects of particle collisions really account for the relationship between size and shape of meshes? Does the "particle size" hypothesis make any predictions that distinguish it from the "resistance to breakage" hypothesis? Formal analysis may help resolve these kinds of questions.

The next step in interpreting design, then, is using models to test the hypotheses. Since both the "resistance" and "particle size" hypotheses might explain the relationship between water velocity and mesh size, a more subtle analysis is needed to distinguish between the following four possibilities: (1) only the "resistance" hypothesis applies, (2) only the "particle size" hypothesis applies, (3) both hypotheses apply, and (4) neither hypothesis applies. A formal analysis might, by generating quantitative predictions from each hypothesis, allow us to distinguish between them. If a model based purely on the "resistance to breakage" hypothesis — incorporating information about silk strength and hydrodynamic forces— accounted for the relationship between current velocity and mesh size, we might tentatively conclude that the essence of the design problem had been captured by these factors alone. We would then have to develop models of the alternative hypotheses to see if they could also account for the data. If more than one model accounted for the data, even in quantitative detail, then even formal modeling would lead to an ambiguous result.

1.2 The Elements of Foraging Models

The foraging models we describe in this book, and optimality models in general, are made up of three components.

1. *Decision Assumptions.* Which of the forager's problems (or choices) are to be analyzed?
2. *Currency Assumptions.* How are various choices to be evaluated?
3. *Constraint Assumptions.* What limits the animal's feasible choices, and what limits the pay-off (currency) that may be obtained?

These components may not always represent mutually independent parts of the problem; for example, constraint assumptions clearly depend on what is being constrained. Some authors have broken foraging models into different components (Schoener 1971, Cheverton et al. 1985, Kacelnik and Cuthill 1986).

Chapter 2 discusses the so-called "conventional models" of foraging theory, and the following chapters discuss changes in the conventional constraint, currency, and decision assumptions in turn (constraints, Chapters 3 and 4; currencies, Chapter 5 and 6; decisions, Chapter 7). The remainder of this chapter presents some general comments about each of these three elements.

1.3 Decision Assumptions

All optimality models consider the "best" way to make a particular decision. How should bones be constructed? What mesh size should a net-spinning caddisfly choose? *Decision* here refers to the type of choice (mesh size or mesh shape?) the animal is assumed to make (or that natural selection has made for it), rather than a specific choice (i.e. deciding that mesh size is to be 10.0 microns by 11.5 microns).

In a formal model the decision studied must be expressed as an algebraic variable (or variables). Mesh size and shape can be represented by a pair of numbers, length and width. Length and width are the *decision variables*. It will not always be possible or reasonable to express the decision as one or two simple variables. A complicated vector (or list) of many decision variables may be more appropriate; for example, in a model of bone structure the *decision vector* might include variables representing bone length, cross-sectional area, alignment of bone fibers, and the locations of muscle attachments.

For some problems even a huge list of decision variables may not suffice. Suppose that a caddisfly larva's body size affects the structure of its net, and, because net structure partially determines the amount of food captured today, that the structure of today's net in turn affects the caddisfly larva's body size tomorrow. A problem like this one is *dynamic*, because today's decision (net structure) affects tomorrow's state (body size), which may in turn affect tomorrow's decision. Dynamic models solve for the optimal path or sequence of decisions. When the decision can be represented by a simple, non-sequential list of decisions, the model is a *static* model. Most of the models we will examine are static, but dynamic models are discussed in Chapter 7.

Foraging models have studied two basic problems: which prey items to consume and when to leave a patch. Modelers have represented the decision variables within each category in different ways. For example, with regard to the first problem, some models of diet have studied the proportions of food of a given type ingested, and others have studied the probability of pursuing a given prey type upon encounter. These diet models, which make different decision assumptions, are studying different aspects of the problem.

Most models of diet choice solve for the optimal *probability that the forager will pursue a given prey type after encountering it*. Two essential ideas are implicit in this assumption: encountering and recognizing prey types. The notion of "prey type" shows how the three components of a model can be interrelated, because the forager's ability to categorize its prey into types (a constraint assumption) is implied by this decision assumption.

In most models of patch exploitation the decision variable is *time spent in a particular patch type* or, more simply patch residence time. This decision variable combines mathematical convenience (time is a convenient and continuous variable) and generality. In most examples of animals exploiting patches, whether the patches are clumps of grass, seeds on a tree, or schools of fish, the forager's decision can be framed in terms of time spent in the patch. However, this assumption can be misleading when there is no strong link between patch residence time and the amount of food acquired from the patch (Chapter 4).

1.4 Currency Assumptions

A model's currency is the criterion used to compare alternative values of the decision variable. A modeler might compare alternative designs of caddisfly nets using a model that assumes maximization of the number of particles filtered per minute. In general, the modeler supposes that trait X will exist instead of other traits if X satisfies some existence criterion. Existence criteria have two parts: a *currency* and a *choice principle*. For caddisfly nets the currency is the "number of particles filtered per minute" and the choice principle is "maximization."

Currencies are as diverse as the adaptations they are used to study, but there are only three common choice principles: maximization, minimization, and stability. Stability is the most general of these, but its generality is not always necessary. If the pay-off (currency) gained by implementing decision X depends on the decisions made by other individuals, then stability is the correct choice principle (Maynard Smith 1982). The models in this book all use maximization or minimization, and so they require that a decision's value is independent of its frequency. (We usually refer only to maximization when speaking in general terms, because any minimization problem can be restated as a maximization problem by maximizing negative currency.) Once the currency and decision variables have been chosen, the modeler must specify the relationship between them.

For example, a modeler might deduce the relationship between the number of particles captured per minute (the currency) and the dimensions of a mesh (the decision variables). This *currency function* must translate the list of decision variables into a single value (in mathematical jargon it must be a real-valued function), because the currency function must rank all possible decisions. (Chapters 5 and 6 discuss ranking alternatives in more detail.)

In most biological optimization problems the modeler chooses a currency *a priori*, largely on the basis of intuition; for instance, a modeler may argue that maximizing the number of particles trapped per minute will make the "fittest" caddisfly, because food limits larval growth. *A priori* currencies usually have physical interpretations: they can be expressed as rates or amounts, for example. Economists and psychologists, on the other hand, often use the observed behavior of a "decision-maker" to specify the currency *a posteriori*. A modeler might suppose that a currency for caddisfly nets has the general form maximize $a\ell^2 + bw^2$, where ℓ and w are the decision variables mesh length and width. An advocate of *a posteriori* modeling would fit the constants a and b so that observed net structure maximized this function. *A posteriori* currencies do not usually have any physical interpretation: they are simply "that which is maximized." Houston et al. (1982) refer to the distinction between *a priori* and *a posteriori* modeling as the distinction between normative and descriptive optimization modeling, and Maynard Smith (1978) refers to the second approach as "reverse" (or "inverse"—McFarland and Houston 1981) optimization. Most of the models we discuss suppose *a priori* currencies, but Chapters 5 and 6 discuss *a posteriori* currencies.

Conventional foraging models maximize the net rate of energy gain while foraging. More energy is assumed to be better, because a forager with more energy will be more likely to meet its metabolic requirements, and it will be able to spend spare energy on important non-feeding activities such as fighting, fleeing, and reproducing. Energy can be measured both as a cost (the energy expended in performing a particular behavior) and as a benefit (the energy gained by performing a particular behavior). Thus it is possible to talk about the net energy gained from performing a particular foraging behavior. Time is critical because animals may be pressed to meet their daily feeding requirements, and because animals are assumed to fight, flee, and reproduce less well if they are simultaneously foraging.

Schoener (1971) pointed out that there are two simple ways to resolve the dilemma of how to acquire more food while spending less time foraging. The *time minimizer* minimizes the time required to gain a fixed ration of energy. The *energy maximizer* maximizes the amount of energy gained in a fixed time. Both alternatives are plausible currencies, but for many purposes both currencies are equivalent to rate maximization (Pyke et al. 1977). These currencies can differ from each other when food comes in lumps. Suppose you can eat from one of two boxes of food: box A contains many items that yield 8 calories and take 3 seconds to eat; box B contains many items that yield 9 calories and take 4 seconds to eat. Whether the currencies agree or disagree depends on what happens if you do not have time to eat a whole item. If the proportion of total calories you take in is the same as the proportion of the required time you spend eating (e.g. you take in one-half the calories if you spend one-half the time), then all three currencies make the same prediction: choose box A because $\frac{8}{3} > \frac{9}{4}$. However, if you do not take in any calories unless you spend all the required time (e.g. you cannot eat until you crack the nut) the currencies are different. Specifically, if you maximize energy gains in 4 seconds, then box B is the better choice—9 calories instead of 8—but if you minimize the time to take in 5 calories, then box A is better—3 seconds to take in 8 calories is better than 4 seconds to take in 9 calories because any amount above 5 calories is sufficient.

In a stochastic world (a world with random variation) foraging theorists must use averages (or expectations) to characterize rates. However, the value of an average rate calculated over 10 seconds may be different from an average rate calculated over 20 seconds: which average is best? Conventional theory has favored generality and mathematical convenience by maximizing the long-term average rate of energy intake (see Box 2.1).

There are still those (e.g. Tinbergen 1981) who confuse maximizing net rate with maximizing the ratio of benefit to cost, often called "efficiency." Although there are conditions under which maximizing efficiency makes sense (for example, allocating resources from a fixed total budget—Schmid-Hempel et al. 1985), for most of the foraging problems we discuss it does not. It ignores the time required to harvest resources, and it fails to distinguish between tiny gains made at a small cost and larger gains made at a larger cost: for example, 0.01 calories gained at a cost of 0.001 gives the same benefit/cost ratio as a gain of 10 calories costing 1. The 10-calorie alternative, however, yields 1000 times the net profit of the 0.01 alternative.

1.5 Constraint Assumptions

By *constraints*, we mean all those factors that limit and define the relationship between the currency and the decision variable(s). This is a broad definition that encompasses both the mathematician's formal use of *constraint* and the everyday use.

A mathematician might define constraints in a purely formal way. Suppose that a currency function relates the number of particles a caddisfly net intercepts per minute (P) to the decision variables length (ℓ) and width (w) and to the stream velocity (v). A modeler might write the function as $P(\ell, w, v)$, and might specify a formal constraint, for example, that the mesh area ℓw must be less than 10 square microns. In the mathematician's purely formal sense the inequality $\ell w < 10$ square microns would be the only constraint in the problem. However, in the everyday sense we would say that the stream velocity constrains the caddisfly's economy.

In the everyday use of "constraint" we imagine some kind of limitation. Limitations are of two biologically different types: those that are intrinsic to the animal and those that are extrinsic. Intrinsic constraints can be further divided into the following categories: (1) limitations in the abilities of animals: honeybees cannot distinguish red from grey, and pigeons cannot distinguish 0.2 milliseconds from 0.5 milliseconds; (2) limitations in the *tolerances* within which animals must live: the forager must acquire 20 milligrams of vitamin A, or it can only tolerate 2 hours of food deprivation. Some biologists have limitations on abilities in mind when they discuss constraints (Janetos and Cole 1981), but others are imagining tolerances (Pulliam 1975).

Extrinsic constraints are placed on the animal by the environment. For example, the stream velocity limits the number of particles that a caddisfly net can filter per hour, and a forager cannot eat more prey than it can find or spend more than 24 hours eating each day. Intrinsic and extrinsic constraints are not mutually exclusive categories. Animal abilities interact with the environment; for instance, both ambient temperature and muscle physiology limit a lizard's running speed. In Chapter 8 we will return to the subject of intrinsic constraints, discussing rules that animals with limited abilities might use to solve their foraging problems.

Conventional foraging models have assumed few constraints on foragers' abilities, and in some important cases they have assumed "no constraints." In our terms even the assumption of "no constraints" is a constraint assumption: it is an assumption about the nature of the limitations on the forager. The advantage of making few constraint assumptions stems from the fact that the limitations on animal abilities vary greatly from species to species: snails and ospreys are not limited in the same ways. Foraging theorists have tried to find general design principles that apply regardless of the mechanisms used to implement them. For example, the elementary principles of a device for getting traffic across a river— that is, a bridge—apply regardless of whether the bridge in question is built of rope, wood, concrete, or steel.

Conventional foraging models make three constraint assumptions: (1)

exclusivity of search and exploitation: the predator cannot exploit (handle) items such as prey or patches while searching for new ones; (2) *sequential Poisson encounters:* items are encountered one at a time, and the probability of encountering each prey or patch type in a short time period is constant; and (3) *complete information:* the forager knows, or behaves as if it knows, the rules of the model. The rules of the model will usually include information about the environment (density of prey) and limitations on the forager's ability. We call this complete information rather than perfect information because it does not imply that the forager is omniscient. A completely informed forager is like a gambler who knows the odds but cannot predict exactly what number will come up on the next spin of the wheel. The assumption of complete information is justifiable for predators in steady-state conditions. Foraging theory has not ignored the question of information gain (see Chapter 4), but the simpler steady-state models are an easier and more useful starting point (Staddon 1983).

1.6 Lost Opportunity

Perhaps the two most important assumptions of the conventional foraging models are long-term average-rate maximization (or "rate maximization" for brevity) and the exclusivity of searching and exploiting. Combining these two assumptions leads to what might be called the *principle of lost opportunity.* In general terms decisions about exploiting items can be assessed by comparing potential gains from exploitation with the potential loss of opportunity to do better. For example, if an item is of the best possible type, then no opportunity can be lost by eating it, since the best outcome that might result from "not eating" it is to happen immediately upon another item of the best sort. By the reverse argument, a forager loses some opportunity when it attacks an inferior item. Many of the results of rate-maximizing theory can be viewed in this way. Gains are assessed in terms of immediate achievements of rate, but losses are assessed in terms of missed opportunities to do better.

1.7 Summary

Formal models of design are valuable because they permit both rigorous analysis and testing. Optimization models consist of three components— decision assumptions, currency assumptions, and constraint assumptions.

The decisions studied by conventional foraging models relate to prey choice and patch exploitation; the currency in these models is long-term average-rate maximization; and the constraints are exclusivity of search and exploitation; sequential, random search; and the assumption of complete information. Many of the results of conventional foraging models are expressions of the principle of lost opportunity.

2

Average-Rate Maximizing:
The Prey and Patch Models

MacArthur and Pianka (1966) distinguished exploiting prey from exploiting patches. Following MacArthur and Pianka's definitions, foraging theorists usually think of prey as discrete items that a forager captures and completely consumes, but they think of patches as clumps of food or simply heterogeneities in the prey distribution. These two ideas stimulated the parallel development of two average-rate maximizing models: the prey and patch models. The basic prey and patch models have not only stimulated a considerable amount of data collection, but they have also formed the basis for a large family of average-rate-maximizing models (Chapters 3 and 4). This chapter introduces the prey and patch models. We discuss the two basic models together on a general level, pointing out what they have in common and how they differ, before going on to deal with each model in detail.

2.1 Some General Comments

Encounters and decisions. Both models assume that the forager encounters prey items or patches one after the other (*sequential encounter*), and they assume that foraging consists of many repetitions of the following sequence: search—encounter— decide. *Search* is somewhat of a misnomer for non-encounter; it may be either waiting or active searching (but see section 3.3). A forager "searches" as long as no prey item or patch is detected while foraging. When a forager, using its senses, detects an item, searching stops and an *encounter* occurs. Thus the sensory abilities of the forager determine what constitutes an encounter, and the models predict how items should be treated upon encounter. Ideally, an experimenter should know enough about the forager's sensory abilities to control or independently assess encounters—it is not enough to measure abundance.

Both models use the ideas of search and encounter, but the form of the *decision* taken upon encounter is different. They prey model asks whether a forager should attack the item it has just encountered or pass it over. The prey question is, attack or continue searching? A prey item provides a fixed mean amount of energy, and a fixed mean amount of time is required to pursue, capture, and consume it (taken together, these activities constitute *handling time*). The patch model asks how long the forager should hunt in the patch encountered. The patch question is, how long to stay in a patch? The models presented in this chapter consider how the forager can best make the prey and patch decisions, with "best" meaning that which maximizes the long-term average rate of energy intake.

The two models differ, principally because they analyze different decisions: eat or search versus how long to stay. Viewed in this way, the models provide surprising definitions of prey and patch. A *prey* yields a fixed amount of energy and requires a fixed amount of time to handle: the forager can control neither the energy gained from nor the time spent in attacking a prey item. However, the forager controls the time spent in, and hence the energy gained from, a *patch*, because there is a well-defined relationship between time spent and energy gained. Any predator that sucks the juices out of its prey might be thought of as preying upon patches (Cook and Cockrell 1978, Sih 1980, Lucas 1985). If these definitions were taken strictly, "true prey" would be rare in nature, since rarely is handling time absolutely fixed. However, as we will show later (section 2.4), these definitions do not have to be taken too seriously. Although we call these prey and patch models, these definitions should emphasize that the fundamental difference between the models is the decisions they analyze, not the nature of the items encountered.

Holling's disc equation. Both models use Holling's disc equation (Holling 1959) to obtain the average rate of energy intake. Holling's disc equation assumes that searching and handling are mutually exclusive activities, and that the expected number of encounters is a linear function of time spent searching. These assumptions make the algebra simple.

Let T_s be all the time spent searching and T_h be all the time spent handling; thus $T_s + T_h = T_f$, where T_f is all the time spent foraging. Let E_f be the net amount of energy gained in T_f. The rate (R) we wish to maximize is

$$R = \frac{E_f}{T_s + T_h}. \tag{2.1}$$

Now because encounters are linearly related to T_s, we can express both E_f and T_h as linear functions of T_s. If λ is the rate of encounter with items (λ has units of prey/time or patches/time), then λT_s is the number of prey items encountered. If s is the cost of search per unit time, then sT_s is the total cost of search. We represent the average energy gained per encounter by $\bar{e}$ ($\lambda T_s \bar{e} = E_f$) and the average time spent handling by $\bar{h}$ ($\lambda T_s \bar{h} = T_h$).

By substituting these relationships into expression (2.1), the rate of energy intake becomes

$$R = \frac{\lambda T_s \bar{e} - s T_s}{T_s + \lambda T_s \bar{h}}, \tag{2.2}$$

and T_s is canceled out to give

$$R = \frac{\lambda \bar{e} - s}{1 + \lambda \bar{h}}, \tag{2.3}$$

which is Holling's disc equation. This derivation follows Charnov and Orians (1973).

Stochastic rates. In the section above we have followed the conventional argument for the use of Holling's disc equation in average-rate-maximizing theory. However, these steps involved a shortcut from expression (2.1) to (2.2) which requires some justification. (The lack of this justification in the literature has created some controversy; see Box 2.1, Templeton and Lawlor 1981, Gilliam et al. 1982, Stephens and Charnov 1982, and Turelli et al. 1982). In this section we want to show precisely how expression (2.1) leads to expression (2.2).

Let G_i be the random variable that describes the net gain from the ith item encountered, and let g_i be a particular realization of G_i. Similarly, let T_i be the random variable that describes the amount of time spent searching for and handling the ith item, and let t_i be a realization of T_i. The gains (G_i) and search-handling times (T_i) for the ith encounter may be correlated.

Equation (2.3) equals the expected energy gain per encounter [$E(G)$] divided by the expected time spent (searching plus handling) per encounter [$E(T)$]:

$$\frac{E(G)}{E(T)} = \frac{\lambda \bar{e} - s}{1 + \lambda \bar{h}}. \tag{2.4}$$

However, is $E(G)/E(T) = E_f/T_f$, that is, does the quotient of the expectations equal expression (2.1)? The answer is yes, but it is a conditional yes. The *equality* holds when an infinite number of items is encountered. In symbols:

$$\frac{E_f}{T_f} = \frac{g_1 + g_2 + \cdots + g_n}{t_1 + t_2 + \cdots + t_n} = \frac{\dfrac{1}{n} \sum_{i=1}^{n} g_i}{\dfrac{1}{n} \sum_{i=1}^{n} t_i}, \tag{2.5}$$

BOX 2.1 THE RATE CONTROVERSY

The controversy over the meaning of average rates centers on what Templeton and Lawlor (1981) claim is an ambiguity in the early foraging literature. There is some ambiguity. For example, Pulliam (1974) and Charnov (1976a) sometimes seem to be considering a "rate of energy intake" without making clear what time interval should be used to calculate the rate. However, the body of theory that grew out of the basic models (well before Templeton and Lawlor's complaint—e.g. Oaten 1977, McNair 1979, and Green 1980) often explicitly used long-term average-rate maximization [max $E(G)/E(T)$, where G = gains per encounter and T = time per encounter] and explicitly claimed that such usage followed the basic models. In short, little confusion resulted from earlier ambiguities, at least until Templeton and Lawlor's paper was published.

Templeton and Lawlor argue that early papers can be interpreted to suggest maximization of the expected rate per encounter: max $E(G/T)$. The important question is not, "who said what?" or even, "does it make a difference?" The important question is, what currency makes most sense for foraging animals? The most reasonable view is that there are arguments for and against both currencies. Imagine that a forager can choose between (1) feeding in a patch for 8 minutes and gaining 5 units of food or (2) feeding in an empty patch for 3 minutes and then feeding in a second patch and gaining 6 units of food in 5 minutes. If, as Templeton and Lawlor suggest, the per patch rate $E(G/T)$ is maximized, choice (1) should be selected, since $\frac{5}{8} > [(\frac{0}{3}) + (\frac{6}{5})]/2 = \frac{3}{5}$. This inequality occurs because entering the empty patch lowers the per patch average. However, the long-term rate $E(G)/E(T)$ gives $[(0 + 6)/2]/[(3 + 5)/2] = \frac{6}{8} > \frac{5}{8}$, and a long-term average-rate maximizer would select choice (2). The distinction is that a per patch rate maximizer will not accept a low rate in the present patch to achieve a higher rate later. This shortcoming suggests that long-term average-rate maximization is more realistic, because we usually will want to evaluate the outcome of a series of foraging decisions.

Infinite-rate maximization also has its problems. For example, an infinite-rate-maximizer might go without food for months or years to obtain a sufficiently large eventual gain; in other words, infinite-rate maximization ignores the pattern of food acquisition (Chapter 6).

and the law of large numbers says that

$$\lim_{n \to \infty} \frac{1}{n} \sum_{i=1}^{n} g_i = E(G) \quad \text{and} \quad \lim_{n \to \infty} \frac{1}{n} \sum_{i=1}^{n} t_i = E(T). \tag{2.6}$$

Written on one line this logic is

$$\frac{E_f}{T_f} \to \frac{E(G)}{E(T)} = \frac{\lambda \bar{e} - s}{1 + \lambda \bar{h}} \quad \text{as } n \to \infty.$$

The equality is exact when n is infinite, but it will be a good approximation when n is large (Turelli et al. 1982). Reassuringly, another approach also justifies using Holling's disc equation in the stochastic case. Using renewal theory, Stephens and Charnov (1982) have shown that mean energy gains approach Holling's disc equation asymptotically as foraging time increases.

Recognition and decision. We have already introduced the idea that the prey and patch models assume complete information (Chapter 1). They also assume that a forager recognizes prey and patch types instantaneously, and that the forager's behavior depends on which type it encounters and recognizes. The forager decides beforehand what it will do when it encounters a given prey type: if big, eat; if small, reject. We call this an *encounter-contingent* policy.

2.2 The Prey Model: Search or Eat?

Our presentation of the prey model follows Charnov and Orians (1973). Essentially the same model, or parts of it, has been presented independently by Schoener 1971, Emlen 1973, Maynard Smith 1974, Pulliam 1974, Werner and Hall 1974, Charnov 1976a, and others.

THE MODEL
Assume that searching (the time between encounters) costs s per time unit. Let there be a set of n possible prey types. Four variables characterize each prey type:

$h_i =$ the expected handling time spent with an individual prey item of type i, if it is attacked upon encounter.

$e_i =$ the expected net energy gained from an individual prey item of type i, (if it is attacked upon encounter) plus the cost of search for h_i seconds (sh_i). If $\tilde{e}_i$ is the net gain from an item of type i, then $e_i = \tilde{e}_i + sh_i$. The modified energy value e_i is the difference in gain between eating a type i item (and gaining $\tilde{e}_i$) and ignoring a type i item (and "gaining" $-sh_i$).

$\lambda_i =$ the rate at which the forager encounters items of type i when *searching*.

$p_i =$ the probability that items of type i will be attacked upon encounter (the decision variable).

These variables, together with the assumptions outlined in section 2.1 and the additional assumptions that (1) the time taken to handle an item encountered but not attacked is zero and (2) the net energy gained from an item encountered but not attacked is also zero, allow us to represent the rate of net energy intake as

$$\frac{\sum\limits_{i=1}^{n} p_i \lambda_i e_i}{1 + \sum\limits_{i=1}^{n} p_i \lambda_i h_i} - s.$$

Because s is constant, we can maximize long-term rate by maximizing

$$R = \frac{\sum\limits_{i=1}^{n} p_i \lambda_i e_i}{1 + \sum\limits_{i=1}^{n} p_i \lambda_i h_i}. \tag{2.7}$$

This expression shows the advantage of incorporating search costs into e_i: our derivation can now proceed following conventional derivations of the prey model, as if there were no search costs.

With respect to a given p_i we may rewrite R as

$$R = \frac{p_i \lambda_i e_i + k_i}{c_i + p_i \lambda_i h_i}, \tag{2.8}$$

where k_i is the sum of all terms not involving p_i in the numerator of equation (2.7), and c_i is the sum of all terms not involving p_i in the denominator of equation (2.7). Both k_i and c_i are constant with respect to p_i (Constraint C.3 of Box 2.2). To see how changes in p_i affect R we differentiate

$$\frac{\partial R}{\partial p_i} = \frac{\lambda_i e_i (c_i + p_i \lambda_i h_i) - \lambda_i h_i (p_i \lambda_i e_i + k_i)}{(c_i + p_i \lambda_i h_i)^2}; \tag{2.9a}$$

carrying out the multiplications indicated in the numerator we find that

$$\frac{\partial R}{\partial p_i} = \frac{\lambda_i e_i c_i - \lambda_i h_i k_i}{(c_i + p_i \lambda_i h_i)^2}. \tag{2.9b}$$

Equation (2.9b) shows that the sign of the derivative of R with respect to p_i is independent of the magnitude of p_i. The value of p_i that maximizes the average rate of energy intake must be either the largest feasible p_i ($p_i = 1$) or the smallest ($p_i = 0$). This demonstrates one of the three principal results of the prey model, the *zero-one rule:* a type is either always attacked upon encounter or always ignored upon encounter (see below). Many workers have misunderstood the empirical implications of the zero-one rule (Box 2.3, Stephens 1985).

The next question is what determines whether $p_i = 1$ or $p_i = 0$. The sign of the numerator in equation (2.9b) supplies the answer. The strategy

BOX 2.2 SUMMARY OF THE PREY MODEL

ASSUMPTIONS

Decision
The set of probabilities of attack upon encounter for each prey type, p_i for the ith prey type.
Feasible choices: For all prey types $0 \le p_i \le 1$.

Currency
Maximization of long-term average rate of energy intake.

Constraint
C.1 Searching and handling are mutually exclusive activities: prey are not encountered during handling.
C.2 Encounter with prey is sequential and is a Poisson process.
C.3 The e_i's, h_i's and λ_i's (net energy gain, handling time, and encounter rate for the ith prey type) are fixed and are not functions of p_i.
C.4 Encounter without attack takes no time and causes no change in energy gains or losses.
C.5 "Complete information" is assumed. The forager knows the model's parameters and recognizes prey types, and it does not use information it may acquire while foraging.

IMPORTANT GENERAL POINTS
1. The model solves for an encounter-contingent policy.
2. Diet (e.g. as measured by stomach contents) is not strictly predicted.
3. Food preference (in dichotomous choice situations) is not strictly predicted.
4. Strong tests must have independent measures of encounter.

that maximizes average rate of energy intake is

$$\text{set } p_i = 0 \quad \text{if } e_i c_i - h_i k_i < 0 \text{ or } e_i/h_i < k_i/c_i$$
$$\text{set } p_i = 1 \quad \text{if } e_i c_i - h_i k_i > 0 \text{ or } e_i/h_i > k_i/c_i.$$

(2.10a)

Because there are n prey types there must be n inequalities like those above. To clarify, let us consider the case $n = 2$.

$$p_1 = 0 \quad \text{if } \frac{e_1}{h_1} < \frac{\lambda_2 e_2}{1 + \lambda_2 h_2}$$

(2.10b)

$p_1 = 1$ otherwise.

$$p_2 = 0 \quad \text{if } \frac{e_2}{h_2} < \frac{\lambda_1 e_1}{1 + \lambda_1 h_1}$$

(2.10c)

$p_2 = 1$ otherwise.

BOX 2.3 TESTING THE ZERO-ONE RULE

The zero-one rule (a type should either always be taken or always ignored upon encounter) is the primary result of the prey model, because the other results follow from it. The zero-one rule's implications for empirical tests of the prey model are unclear. We give two examples.

Krebs et al. (1977) tested the prey model by presenting two types of prey (bigs and smalls) to great tits (*Parus major*). The prey moved on a conveyor belt in front of a waiting tit. Krebs et al. varied the encounter rate with the most profitable prey (bigs). The zero-one rule clearly predicts that at some sufficiently high encounter rate with bigs, smalls should *always* be ignored when encountered. Similarly, at some sufficiently low encounter rate with bigs,

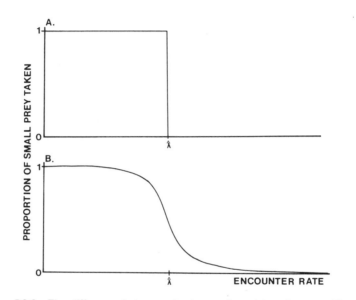

Figure B2.3 The difference between absolute and partial preferences. The encounter rate with the more profitable, large prey (λ) is on the abscissa, and the proportion of the less profitable, small prey attacked is on the ordinate. The prey model's zero-one rule predicts that small prey should be attacked when the encounter rate with more profitable prey is less than some threshold value, $\hat{\lambda}$, and small prey should be ignored when the encounter rate with large prey is greater than the threshold $\hat{\lambda}$. (A) Many workers have interpreted this prediction to mean that absolute preferences should be observed: less profitable prey should be attacked every time they are encountered if $\lambda < \hat{\lambda}$ and ignored every time they are encountered if $\lambda > \hat{\lambda}$. (B) However, if the threshold encounter rate, $\hat{\lambda}$, has some variance, then partial preferences (a smooth, sigmoid relationship) are expected.

BOX 2.3 (CONT.)

smalls should always be taken when encountered. The model predicts a threshold switch in behavior: above some encounter rate with bigs, say $\hat{\lambda}$, smalls ought never to be taken, and below $\hat{\lambda}$, smalls ought always to be taken. The step function in Figure B2.3(A) shows this prediction.

This step-function expectation gives the threshold a trait unheard of in biological measurement: no variance! If there is some variance in the threshold (it fluctuates in a way that the experimenter can only partly control), then the only reasonable expectation is something like the cumulative distribution function of a normal distribution shown in Figure B2.3(B), with mean $\hat{\lambda}$. Using simple but rigorously justified techniques, the experimenter can estimate the mean threshold and then perform statistical tests to discover whether this mean is near the predicted threshold (Finney 1962, Stephens 1985). These techniques skirt the issue of what causes the threshold's variance (see Krebs and McCleery 1984 for a list of possible sources of variation), but this kind of issue-skirting is inherent in statistical tests. Recall that "statistical significance" only measures error given that the underlying hypothesis is correct.

Pulliam (1980) studied seed selection by sparrows (*Spizella passerina arizonae*). Using logic similar to the step-function expectation of Krebs et al., he argued that the least profitable item observed in the diet allowed him to estimate the profitability threshold between acceptable and unacceptable seeds. He therefore claimed that the zero-one rule predicts that all seed "species that are eaten should occur roughly in the same proportions in the diet as they do in the soil." However, if the threshold seed profitability has variance, then neither of these deductions is correct. A "random threshold" model, however, predicts that seeds of lower profitability should be taken less frequently than their occurrence in the environment, because random fluctuation in the threshold means that a given species is sometimes in and sometimes out of the "diet." A type's distance from the mean profitability threshold affects how frequently it is in the diet.

Consider the exclusion ($p_1 = 0$) of type 1. The condition above (2.10b) yields (after a little algebra)

$$p_1 = 0 \quad \text{if } e_1 < \lambda_2(e_2 h_1 - e_1 h_2).$$

The variables h_1, h_2, and λ_2 must be positive by definition. The variables e_1 and e_2 are positive for any prey types worth considering: no prey type that yields less than $-s$ per unit handling time ($e_i < 0$ implies that $\tilde{e}_i/h_i < -s$) should be considered because the forager can always achieve a rate of $-s$ by ignoring everything.

Since all the terms in this expression are positive, the minimum requirements for the exclusion of a given type are:

(i) Type 1's can only be excluded if $e_2/h_2 > e_1/h_1$.
(ii) Type 2's can only be excluded if $e_1/h_1 > e_2/h_2$.

Either (i) or (ii) must be true (assuming $e_2/h_2 \neq e_1/h_1$), so without losing generality we can rank the prey types such that $e_1/h_1 > e_2/h_2$. This means that (i) cannot hold, and type 1's cannot be excluded (we choose to call the type with higher e/h type 1). We need only ask the question, should type 2's be included? Inequality (2.10c) provides the answer. This logic suggests an algorithm for finding the "diet" that maximizes the average rate of energy intake:

Prey Algorithm: Rank the n prey types such that $e_1/h_1 > e_2/h_2 > \cdots > e_n/h_n$. Add the types to the "diet" in order of increasing rank until

$$\frac{\sum_{i=1}^{j} \lambda_i e_i}{1 + \sum_{i=1}^{j} \lambda_i h_i} > \frac{e_{j+1}}{h_{j+1}}. \tag{2.11}$$

The highest j that satisfies expression (2.11) is the lowest ranking prey type in the "diet." If this inequality obtains for no $j < n$, then take all n items upon encounter.

The argument for the $n = 2$ case shows that this algorithm applies in the simplest non-trivial case. Assume that the algorithm is known to produce the maximum average rate of energy intake for any set of k prey types. Does this assumption imply that the algorithm can be applied to a set of $k + 1$ types? Consider an arbitrary set of $k + 1$ types. Rank the set such that $e_1/h_1 > e_2/h_2 > \cdots > e_k/h_k > e_{k+1}/h_{k+1}$. Now consider the ordered subset of this set that excludes only the $k + 1$st item. We know that the algorithm can be applied to this subset of k types. There are two cases. First, if the kth item is in the diet chosen from among the set of items 1 through k, then, applying (2.10a), $k + 1$ should be excluded if

$$\frac{\sum_{i=1}^{k} \lambda_i e_i}{1 + \sum_{i=1}^{k} \lambda_i h_i} > \frac{e_{k+1}}{h_{k+1}}, \tag{2.12}$$

according to expression (2.10). This agrees with the algorithm.

In the second case the kth item is not in the average-rate maximizing diet. Since the algorithm applies in the kth case, there is some $m < k$, where m is the lowest-ranked prey type in the diet. By definition of our ranking,

$$\frac{\sum\limits_{i=1}^{m} \lambda_i e_i}{1 + \sum\limits_{i=1}^{m} \lambda_i h_i} > \frac{e_k}{h_k} > \frac{e_{k+1}}{h_{k+1}}. \tag{2.13}$$

Applying expression (2.10a) shows that the $k + 1$st item cannot be in the rate-maximizing "diet." This inductive argument, together with the two-prey-type case, proves the prey algorithm, since it shows that the algorithm must apply for a set of $k + 1$ prey types if it applies for a set of k prey types.

PREDICTIONS

Listed below are the three principal results of the prey model. The algorithm suggests both the second and third results.

1. *The Zero-One Rule.* Types are either always taken upon encounter ($p_i = 1$) or never taken upon encounter ($p_i = 0$).

2. *Ranking by Prey Profitability.* Types are ranked by the ratio of energy per attack to the handling time per attack. This ratio is called the *profitability* of a prey type. Types are added to the diet in order of their ranks, by the algorithm above (equation [2.11]).

3. *Independence of Inclusion from Encounter Rate.* The inclusion of a prey type depends only on its profitability and on the characteristics of types of higher rank (one is the highest). The inclusion of a type does not depend on its own encounter rate. Equation (2.11) predicts when a type should be attacked.

The last prediction is often thought to be the most surprising of the three. However, if we recall two aspects of the model, encounter-contingent policy making and the principle of lost opportunity (section 1.6), then this prediction is not surprising. The model asks whether a prey type should be attacked after it has been encountered. No opportunity can be lost by attacking an item of the highest possible rank, since the best alternative outcome is immediately to encounter another top-ranking item. For a low-ranking item, the lost opportunity is the expected gain from searching for and eating a higher-ranked item. Equation (2.10) states that if the opportunity loss due to attacking a low-ranking type exceeds the immediate gains from the attack, then it never pays to eat the low-ranking item, regardless of how often it is encountered. The unimportance of a type's own encounter rate seems surprising because many people think of "diet choice" as the problem of deciding what to search for, but this model asks what should be ignored and what should be eaten, given a fixed search method, place, and pattern. Chapter 9 summarizes the empirical evidence bearing on the prey model's predictions.

Diet and Food Preference Not Predicted

The prey model is often called a "diet model," and it is convenient to call "setting the probability of attack equal to one" inclusion in the "diet." However, ecologists have traditionally considered diet to be the set of all food items taken by an animal. The model we have just presented considers diet choice within a homogeneous patch for a forager using a fixed search strategy. If the forager moves to another patch the model should be freshly applied. This means that the model cannot be tested simply by looking, for instance, at stomach contents and taking overall averages to estimate the parameters of the model (the λ_i's, e_i's, and h_i's). For example, a rate-maximizing forager might forage in a part of the environment where low-ranking items are disproportionately common, because the abundance of these items might compensate for their low food value. A forager choosing such a patch would appear to take too many low-ranking types and too few high-ranking types compared with an idealized forager for which we calculate the model parameters as if it foraged in the whole environment. This apparent contradiction might occur even if the forager maximizes its average rate of energy intake both in its choice of where to feed and in the diet chosen there. Because of this problem, criticisms of the prey model based on data such as stomach contents cannot be taken at face value (see Schluter 1981).

The prey model is sometimes used to claim that if the profitability of prey type X is greater than the profitability of prey type Y (i.e. if $e_x/h_x > e_y/h_y$), then type X should be preferred to type Y. "Preference" suggests that X and Y are alternatives. Type X is preferred to type Y if X is chosen when X and Y are offered *simultaneously*. In the prey model X and Y are offered sequentially, and they are not alternatives. Strictly speaking, the ranking of prey types is not a preference ranking. The model does deal with preference in another way, however. If $p_i = 1$ maximizes the average rate of energy intake, then attacking a type i item is preferred to searching further, or if $p_i = 0$ maximizes the average rate of energy intake, then searching further is preferred to attacking. We discuss preference in the strict sense, that is, when a forager encounters items simultaneously, in the next chapter.

2.3 The Patch Model: How Long to Stay?

The model presented here is that of Charnov (1976b: the so-called marginal-value theorem), but we will also discuss related results for cases not explicitly treated by Charnov. To emphasize the similarity between the prey and patch models, our presentation of the marginal-value theorem parallels our presentation of the prey model.

THE CHARACTERISTICS OF A PATCH

Assume that there are n patch types, and that traveling between patches costs s per time unit. Three quantities characterize each patch type:

$\lambda_i =$ the encounter rate with patches of type i.

$t_i =$ the time spent hunting in patches of type i. t_i is called the *patch residence time*, and t_i is the decision variable.

$g_i(t_i) =$ the *gain function* for patches of type i. This function specifies the expected net energy gain from a patch of type i if t_i units of time are spent hunting in each type i patch entered. In statistical notation $g_i(t_i) = E(\text{Gain} \mid T = t_i)$. (Following convention, we do not incorporate search costs, s, in the gain function in the way that we incorporated them into e_i in the prey model.)

NATURE OF THE GAIN FUNCTION

The gain function is assumed to be a well-defined, continuous, deterministic, and negatively accelerated (curving down) function. However, functions may be negatively accelerated in many ways. Figure 2.1 shows six hypothetical shapes for gain functions. If we require that the gain function be negatively accelerated for all residence times (zero to infinity), then only types (A) and (B) are allowable gain functions. However, there is no reason to invalidate sigmoid functions like example (C). We imagine here that gain functions have the following characteristics.

1. The net energy gain when zero time is spent in a patch is zero $[g_i(0) = 0]$.
2. The function is at least initially increasing $[g_i'(0) > 0]$.
3. The gain function is eventually negatively accelerated [there exists some $\tilde{t}$ such that $g_i''(t) < 0$, for all $t \geq \tilde{t}$].

Many workers consider the simple exponential gain function shown in Figure 2.1(A) to be a general description of within-patch gains. However, the humped gain function in Figure 2.1(B) is probably a more general shape for net gains because as the patch becomes depleted more energy will eventually be spent than gained. If gross intake follows the function in Figure 2.1(A), then net intake will approximate Figure 2.1(B).

Depletion and depression. The patch model depends on the assumption that the gain function eventually curves down (shows negative acceleration). Foraging theorists have argued that this must be a general phenomenon because most patches contain finite resources, and thus foraging must deplete them. However, a patch can deplete without having a negatively accelerated gain function. If a forager searched patches systematically, and the prey items were randomly distributed in patches, the

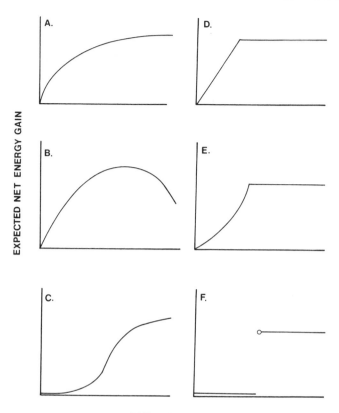

PATCH RESIDENCE TIME

Figure 2.1 Six hypothetical gain functions. (A) A simple exponential gain function that increases asymptotically to a maximum value. This is probably a good description of gross gains for many patches. (B) A humped gain function is probably the most reasonable general description of net foraging gains. (C) A sigmoid gain function might occur in some patches, for example, mobile prey might be able to avoid the forager at first, but the gain rate increases as clusters of prey are discovered. (D) A gain function for a patch that depletes but does not show patch depression. This can occur if the forager searches the patch systematically. (E) A pathological gain function that describes a patch where gains show positive acceleration up to a maximum. (F) The prey gain function: prey theory effectively supposes that prey items show a gain-to-time relationship like this; up to a certain time no energy is gained, and after this time the maximum amount of energy is acquired. Functions (A), (B), and (C) meet the criteria of the marginal-value theorem. According to the combined prey and patch model of section 2.4, functions (D), (E), and (F) can all be treated as if they were "true prey."

gain function would look like Figure 2.1(D); there would be no negative acceleration, even though the patch was being depleted as the forager hunted in it. To distinguish patch depletion from negative acceleration of the gain function, Charnov et al. (1976) coined the phrase patch *depression* to mean a decrease in the instantaneous rate of energy gain within a patch (i.e. negative acceleration of the gain function). Charnov et al. also discuss ways in which patch depression might arise (for example, random searching in patches being depleted and the evasive behavior of prey can cause depression) and its probable generality.

We assume for now that the decision about which patches should be entered (which patches should be included in the "diet": the prey question) has already been made. This may not be reasonable, because the value (i.e. the profitability) of the patches in question will change as we change the t_i's (the decision variables).

The assumptions above (see Box 2.4 for a complete list) allow us to specify the average rate of energy intake (R) as

$$R = \frac{\sum_{i=1}^{n} \lambda_i g_i(t_i) - s}{1 + \sum_{i=1}^{n} \lambda_i t_i}. \tag{2.14}$$

We want to maximize R by choosing the set of t_i's. [In technical jargon we wish to choose the optimal vector $(t_1, t_2, \ldots, t_n)$]. We differentiate with respect to a given t_i:

$$R = \frac{\lambda_i g_i(t_i) + k_i}{c_i + \lambda_i t_i} \tag{2.15}$$

$$\frac{\partial R}{\partial t_i} = \frac{\lambda_i g_i'(t_i)[\lambda_i t_i + c_i] - \lambda_i[\lambda_i g_i(t_i) + k_i]}{(\lambda_i t_i + c_i)^2}. \tag{2.16}$$

As in the prey model, k_i is the sum of all terms not involving t_i in the numerator, and c_i is the sum of all terms not involving t_i in the denominator. It can be shown that R is maximized where $R'(t_i) = 0$ (see Box 2.5):

$$g_i'(t_i)[\lambda_i t_i + c_i] - [\lambda_i g_i(t_i) + k_i] = 0 \tag{2.17}$$

$$g_i'(t_i) = \frac{\lambda_i g_i(t_i) + k_i}{\lambda_i t_i + c_i}. \tag{2.18}$$

Notice that the right side of this condition is the average rate of energy intake and the left side is the instantaneous rate of gain within a patch

---

BOX 2.4 SUMMARY OF THE PATCH MODEL

ASSUMPTIONS

Decision

   The set of residence times for each patch type, t_i for patch type i.
   Feasible choices: For all patch types $0 \le t_i < \infty$.

Currency

   Maximization of long-term average rate of energy intake.

Constraint

C.1 Searching for and hunting within patches are mutually exclusive activities.

C.2 Encounter with patches is sequential and is a Poisson process.

C.3 Encounter rates when searching are independent of the residence times chosen.

C.4 Net expected energy gain in a patch is related to residence time by a well-defined gain function $[g_i(t_i)]$ with the following characteristics:
   (i) Change in energy gain is zero when zero time is spent in a patch.
   (ii) The function is initially increasing and eventually negatively accelerated.

C.5 Complete information is assumed. The forager knows the model's parameters and recognizes patch types, and it does not acquire and use information about patches while foraging in them.

IMPORTANT GENERAL POINTS

1. The model solves for an encounter-contingent policy.

2. Significant changes must be made to the model if the forager assesses patch quality while hunting in patches.

3. The model applies only to patches with negatively accelerated gain functions—patch depression. This should be confirmed by observation in empirical tests of the model.

4. The marginal-value condition, equation (2.18), gives only an implicit solution of the rate-maximizing patch residence time. It is incorrect to treat the average rate of energy intake, the right side of equation (2.18), as if it were independent of patch residence time.

---

of type i, at time t_i. Since we are solving for all n t_i's, we have a set of n equations in n unknowns:

$$
\begin{aligned}
g'(\hat{t}_1) &= R(\hat{t}_1, \hat{t}_2, \dots, \hat{t}_n) \\
g'(\hat{t}_2) &= R(\hat{t}_1, \hat{t}_2, \dots, \hat{t}_n) \\
&\vdots \\
g'(\hat{t}_n) &= R(\hat{t}_1, \hat{t}_2, \dots, \hat{t}_n).
\end{aligned}
\tag{2.19}
$$

BOX 2.5 THE MARGINAL-VALUE THEOREM:
THE SECOND ORDER CONDITION FOR MAXIMIZATION

We consider only the simplest—one patch type, nil search costs—case. Here, the derivative of the long-term average rate of energy intake with respect to patch residence time is

$$R'(t) = \frac{\lambda g'(t)[1 + \lambda t] - \lambda^2 g(t)}{(1 + \lambda t)^2} = \frac{N(t)}{D(t)}.$$

Let $N(t)$ be the numerator of this expression and $D(t)$ be the denominator. We wish to prove that when $N(\hat{t}) = 0$, R is at a maximum. To do this, we need to know the sign of the second derivative of R with respect to t.

$$R''(t) = \frac{N'(t)D(t) - D'(t)N(t)}{[D(t)]^2}$$

but at $\hat{t}$, $N(\hat{t}) = 0$. So,

$$R''(\hat{t}) = \frac{N'(\hat{t})D(\hat{t})}{[D(\hat{t})]^2} = \frac{N'(\hat{t})}{D(\hat{t})}$$

and, substituting the definitions of $N(t)$ and $D(t)$ back into this expression,

$$R''(\hat{t}) = \frac{\lambda g''(\hat{t})}{1 + \lambda \hat{t}}.$$

Therefore, $R''(\hat{t})$ has the same sign as $g''(\hat{t})$. Since we have assumed that $g''(\hat{t})$ is negative, $R''(\hat{t})$ must be negative, and $R(\hat{t})$ must be a maximum. The reverse is also true: if $g''(\hat{t})$ is positive (positively accelerated), then the marginal-value condition gives a local minimum.

This set of equations gives a condition that the rate-maximizing set of t_i's must fulfill. The condition is Charnov's marginal-value theorem: a rate-maximizing forager will choose the residence time for each patch type so that the *marginal* rate of gain at the time of leaving equals the long-term average rate of energy intake in the habitat. The phrase "marginal rate" comes from economics where it translates as "derivative."

The set of equations above gives two results. First, the marginal rate at leaving must be the same in all patches visited. Second, if the habitat becomes poorer (if the average rate of energy intake decreases) without affecting the gain function of type i patches, then a rate-maximizing forager will stay longer in type i patches (equivalently, a rate-maximizer will leave patches earlier if the average rate of gain increases). The habitat may become poorer without affecting the gain function of a given patch type either because the encounter rates decrease (patches become farther apart, so that the forager spends more time traveling between patches) or because the set of patch types visited is reduced.

The marginal-value theorem is most easily understood when all patches have the same gain function and search costs are nil ($s = 0$). Figure 2.2 shows the well-known graphical solution of this case. In the one-patch-type case the condition above, equation (2.18), becomes

$$g'(\hat{t}) = R(\hat{t}) = \frac{\lambda g(\hat{t})}{1 + \lambda \hat{t}}. \tag{2.20}$$

Many people immediately think of a simple graph like Figure 2.2 (with the tangent "rooted" at $1/\lambda$ on the travel time axis) when they think of the marginal-value theorem, and in this chapter and the next we use this formulation to analyze patch-use problems qualitatively. However, it is important to remember that this is a special case of the marginal-value theorem that requires some restrictive assumptions (one patch type, $s = 0$). This graphical solution can also be used if the experimenter plots gross energy gains on the ordinate (instead of net gains as we have assumed) and if between- and within-patch energy expenditures are the same per time unit (see section 9.4, Kacelnik and Houston 1984). Notice that the time between encounters is usually called "travel time" in the patch model, although it is often called "search time" in the prey model.

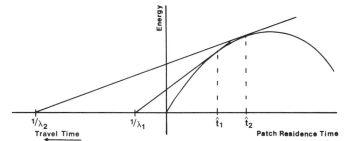

Figure 2.2 The marginal-value theorem in the one-patch-type case. This graph is unusual because we plot two quantities on the abscissa: travel time increases from the origin to the viewer's left, and patch residence time increases from the origin to the viewer's right. The optimal residence time can be found by constructing a line tangent to the gain function that begins at the point $1/\lambda$ on the travel time axis. The slope of this line is the long-term average rate of energy intake, because $1/\lambda$ is the average time required to travel between patches. When the travel time is long $(1/\lambda_2)$, then the rate-maximizing residence time $(\hat{t}_2)$ is long. When the travel time is short $(1/\lambda_1)$, then the rate-maximizing residence time $(\hat{t}_1)$ is shorter.

Implicit solutions. Patch residence time $\hat{t}$ appears on both sides of equations (2.18), (2.19), and (2.20). This means that the marginal-value theorem only gives an implicit solution of the rate-maximizing residence time (or times). If the gain function is complicated, we may have to evaluate this expression numerically to find $\hat{t}$. Some authors (e.g. Pyke 1978a) have treated the average rate, the right side of equations (2.18), (2.19), and (2.20), as a constant and measurable feature of the environment. According to this view, the the rate-maximizing residence time can be determined by setting the marginal rate of gain equal to the "constant" rate and solving this much simpler equation: $g'(t) =$ a constant. This is wrong, because patch residence time affects both the marginal rate of gain and the average rate of energy intake. To show that these numbers are similar only weakly supports the marginal-value theorem, because this equality is a necessary but not sufficient condition for rate maximization. Strong tests will have to solve equation (2.18) for $\hat{t}$. For example, a strong test might establish the explicit relationship between the encounter rate (λ) and the rate-maximizing patch residence time ($\hat{t}$), as Cowie (1977) and Giraldeau and Kramer (1982) have done (see Chapter 9 for more examples).

Rules and rates. After their original development of this model, Charnov and Orians (1973) made the following summary statement:

> When the intake rate in any patch drops to the average rate for the habitat, the animal should move on to another patch. Thus, the choice is not really one of t_i, it is the "moving-on threshold" in intake rate that is important.

This statement shows the intuitive beauty of the marginal-value theorem, but it also contains the seeds of two misunderstandings because it is imprecise in two details. First, the phrase "intake rate" is meaningless unless we specify the time over which the intake is measured. The theorem explicitly deals with the instantaneous (or marginal) intake rate. Second, the mathematics we have used to develop the marginal-value theorem does not justify the switch from solving for the optimal residence time to solving for a general "moving-on threshold." It does hint at such a result, but it cannot establish it, because we have shown only that the moving-on threshold follows from the assumption that residence times are chosen by the forager.

The idea of a moving-on threshold implies that the forager continually measures or assesses the instantaneous rate of gain when foraging in a patch and leaves when this "measured instantaneous rate" drops to the long-term average rate of energy intake. Such a procedure violates the assumption of encounter-contingent policy making (i.e. the assumption that the forager must choose the same mean residence time for all patches of type i, every time they are encountered). Oaten (1977), Green (1980), and McNamara (1982) have shown that when conditions make encounter-contingent policies unreasonable (specifically, when foragers use information gained from hunting in a patch to change their tactics within that patch), then a moving-on threshold using a measured marginal capture rate fails to maximize the long-term rate of energy gain (Chapter 4).

2.4 Combining the Prey and Patch Models

The prey and patch models both assume that the decision analyzed by the other has been made: the prey model assumes that handling times and energy gains are outside the forager's control; similarly, the patch model assumes that the set of patch types that will be attacked is outside the forager's control. These assumptions can only be justified if there is a degree of independence between the models. Moreover, we need to know how to combine the two models to build a general average-rate-maximizing theory.

We present a model that treats all items encountered as patches and asks which patches should be entered *and* for how long. We associate with each of the n patch types four things:

$g_i(t_i)$ = the gain function (we relax the assumption of eventual depression to the assumption that the patch holds finite resources). For simplicity, we assume that search costs are nil ($s = 0$).

λ_i = the encounter rate.

t_i = the time spent handling, a decision variable.

p_i = the probability of attack upon encounter, a decision variable.

We wish to find the set of pairs $[(t_1, p_1); (t_2, p_2); \ldots ; (t_n, p_n)]$ that maximizes the average rate of energy intake. It is easy to show (by differentiation) that the zero-one rule still applies, and that a patch should be included in the "diet" if

$$g_i(t_i)/t_i > k_i/c_i$$

(k_i and c_i are as defined in previous sections). However, the interpretation of the quotient on the left is not straightforward because it is a function of the variable t_i. Consider the meaning of this inequality. If it is true, then increasing p_i (to 1) increases the average rate of energy intake. If, for a given patch type, there exists any t_i such that the inequality holds, then the average rate of energy intake can be increased by including that patch type. If the inequality holds for any t_i, it must hold for the t_i that maximizes $g_i(t_i)/t_i$. This argument justifies changing this inequality to:

Include type i if $g_i(t_i^*)/t_i^* = \max_{t_i}[g_i(t_i)/t_i] > k_i/c_i$.

This inequality suggests that patches should be ranked according to their maximum profitabilities. We know from the previous section that the marginal-value theorem yields the maximum average rate of energy intake, given that the "diet" is already determined. The ranking we have deduced (plus the zero-one rule) provides us with an orderly way to work through the possible "diets":

> *"Patches as Prey" Algorithm.* Rank the n patch types according to their maximum profitabilities: $g_1(t_1^*)/t_1^* > g_2(t_2^*)/t_2^* > \cdots > g_n(t_n^*)$. Add types to the diet in order of rank until

$$R = \frac{\sum\limits_{i=1}^{j} \lambda_i g_i(\hat{t}_{ij})}{1 + \sum\limits_{i=1}^{j} \lambda_i \hat{t}_{ij}} > \frac{g_{j+1}(t_{j+1}^*)}{t_{j+1}^*}, \qquad (2.21)$$

where $\hat{t}_{ij}$ indicates the residence time for the ith patch type that maximizes the average rate of energy intake when patch types one through j are included in the "diet." The marginal-value theorem specifies the average rates of energy intake on the left side of inequality (2.21).

An inductive proof of this algorithm is straightforward and similar to the proof of the prey algorithm (see section 2.2, section 3.2, Stephens et al. 1986).

The combined prey-patch model preserves the marginal-value theorem because it is independently applied at each step in the algorithm. The combined model also preserves the prey model's three principal results (zero-one rule, ranking by prey profitability, and independence of inclusion from encounter rate), and its shows that they can be applied to a large class of ecological entities (see section 3.2).

This algorithm raises interesting questions about gain functions. Notice that for any patch, $\hat{t}$ must always be greater than or equal to t^*, and t^* will be equal to $\hat{t}$ for any patch in which gains increase linearly or faster [$g''(t) \geq 0$] to a finite limit. Notice also that $\hat{t}$ will be greater than t^* for any patch with negative acceleration [$g''(t) < 0$]. This result ties up a loose end in the marginal-value theorem. It guarantees that for any patch (with a continuous gain function) that is entered [$g'(t^*) = g(t^*)/t^* > R$], there will exist a point such that the marginal-value condition holds [$g'(\hat{t})$ = average rate of energy intake] (see Box 2.5). Moreover, for an average-rate-maximizing forager, there are only two plausible kinds of patches: (1) those patches for which the maximum profitability of the patch [$g(t^*)/t^*$] occurs at the same residence time as the maximum energy gain, as represented in Figure 2.1(D), (E), and (F): these patches show no patch depression; (2) those patches for which the maximum profitability occurs before (at a smaller t than) the maximum energy gain, as represented in Figure 2.1(A), (B), and (C): these patches show patch depression.

An average-rate maximizer (which does not use information, either because it cannot or need not) should treat a patch without depression exactly like a prey item. The maximum profitability effectively characterizes the gain function, and the patch should be handled (if it is handled at all) for a fixed time regardless of the average rate of energy intake. We must disallow the use of information because we know from Green's (1980) work that if patch sampling occurs, then patches with no depression should be treated differently from prey items (see Chapters 4 and 8).

2.5 Limitations

Although the conventional prey and patch models are the most widely known foraging models, they are only the starting points for further analysis. Here we discuss three of these models' most important shortcomings, each of which will be discussed in later chapters.

Beyond time and energy. Like all economic models, the prey and patch models attempt to combine many factors into a single currency that can be used to evaluate strategies. They do this by expressing the consequences of alternative strategies in terms of the long-term rate of energy intake. However, even the more ardent supporters of average-rate maximization acknowledge that other factors, such as wariness, dominance, and territory defense, often influence foraging behavior. There is no logical difficulty about putting more factors into optimality models when we know, or can guess, how to combine them into common currency. Chapters 5 and 7 address this problem.

Static versus dynamic. A second major criticism is that the basic models are static (as defined in section 1.3). They do not take forager's state into account; for example, they ignore whether the forager is satiateed or starving. This difficulty is addressed in Chapter 7.

Information. As we have explicitly assumed, the basic models do not allow the forager to use (i.e. to change its behavior in response to) information gained while foraging. Pyke et al. (1977) have stated that animals might be viewed as statisticians who estimate the parameters of distributions. This is only partly correct, because in most applications of statistics, statisticians do not worry about the *value* of knowledge. In foraging theory (and in almost any economic models of decision making) some pieces of information have little or no value and are not worth the purchase price. In average-rate-maximizing models the value of a piece of information must be defined in terms of how it changes the average rate of energy intake. Since the forager uses information in the future, assessing the value of information can be a very complicated problem (Chapter 4). Information is sometimes confused with the more general problem of stochasticity: for example, Oaten's (1977) important critique of the marginal-value theorem, although partially entitled "a case for stochasticity," is really about the use of information.

2.6 More Decisions for the Average-Rate Maximizer

HABITAT SELECTION

The patch model supposes that patches are relatively small and quickly depleted. When a forager's behavior does not deplete the resources within a patch, we may think of the resources as being effectively infinite. We call such an "infinite" patch a *habitat:* the forager's behavior changes neither the rate of gain from a habitat, nor the time for which this rate can be maintained.

This definition means that a habitat is completely characterized by its average rate of gain. How should a forager choose among habitats? When it knows the rates of gain, and these rates are constant, it should simply choose the habitat that has the highest average rate of energy intake. Interesting modifications in this selection process occur under conditions of varying rates, imperfect information, differential likelihoods of predation, proximity to the central place (which we discuss in later chapters), and social interactions (Fretwell 1972).

SEARCH METHOD

The prey and patch models assume that certain decisions about search method have already been made, but many foragers have more than one search method available to them: a trout might either lie in wait for drifting insects or patrol a large section of stream bed. Choices of search method can be treated like habitat selection problems if each choice can be characterized by its long-term average rate of energy intake. One qualitative difference between selection of search method and selection of habitat is that the cost of changing from one search method to another will usually be lower than the cost of changing habitats. As a general rule, the increasing costs of switching between search modes or habitats mediate against change. Thus we expect that foragers will change their search method more readily than their habitat.

Interestingly, if the gains from a search method decrease in marginal value with time, then a model similar to the marginal-value theorem could be used to determine how long a search method should be used.

TRAVEL: COSTS, SPEEDS, AND ROUTES

Costs and speed. The basic average-rate-maximizing models assume that encounter rates and search costs are fixed constraints that the forager must abide by. Although this assumption is not completely unreasonable, many foragers will have some latitude in choosing how quickly and how costly their searching will be (see Evans 1982). Search speed and search costs may affect prey and patch decisions (DeBenedictis et al. 1978, Schmid-Hempel et al. 1985). Ware (1975) argues that the most important foraging decision for the pelagic fish *Alburnus* is the choice of an optimal swimming speed. Ware's model treats *Alburnus* as if it were a net being pulled through the water. There is an optimal pulling (swimming) speed because there is a nonlinear relationship between speed and energy expenditure, and there is also a relationship between speed and rate of energy acquisition. Ware's model complements the models presented here, because he assumes that decisions about which items to select and how to treat them have already been made, independent of the choice of swimming speed. It may be unreasonable to treat foraging speed as a factor that is independent of other foraging decisions, but it is not obvious where this independence assumption will fail (see Pyke 1981a).

Routes. The choice of routes might also change the basic models. For example, if the types of prey or patches already encountered affected the forager's search method (e.g. turn left after eating a caterpillar), then this would violate the assumption that selection (p_i) and patch exploitation (t_i) can be changed independent of encounter rates (Constraint C.3 of Box 2.2 and Constraint C.3 of Box 2.4). McNair (1979) has considered the general theoretical implications of violating this assumption; however, little empirical work has addressed this important problem. Pyke (1983) has reviewed the problems of travel from an economic perspective.

2.7 Summary

This chapter presents the basic average-rate-maximizing models: the prey and patch models. These models differ because they analyze different decisions, not because one deals with prey and the other deals with patches.

The prey model comprises three basic results: the zero-one rule, ranking prey by profitability, and the independence of a type's inclusion in the "diet" from its own encounter rate. The patch model's main result is the marginal-value condition: set the patch residence time so that the instantaneous rate of gain at leaving equals the long-term average rate of gain in the habitat.

We also present a combined prey and patch model, in which the forager chooses both the probability of entering a given patch type upon encounter and the patch residence time. This combined model preserves the main features of the separate models.

We emphasize the limitations and difficulties of the basic models. Some particularly important limitations are (1) the assumption of encounter-contingent policy making; (2) the assumption of complete information; (3) the limited sense in which the prey model predicts "diets"; (4) the inability of the prey model to predict preference; (5) the lack of implication, in the patch model, of a rule by which the forager "measures" the instantaneous rate of gain in a patch; and (6) the provision by the patch model of only an implicit solution of the rate-maximizing patch residence time.

3

Average-Rate Maximizing Again: Changed Constraints

3.1 Introduction

This chapter considers long-term average-rate-maximizing foraging models with changed *constraint* assumptions. As we pointed out in Chapter 1, the basic prey and patch models include constraint assumptions that are in effect "unconstraints," such as unrestricted travel and complete information. This chapter adds more realistic constraints to the basic models of Chapter 2. Some of these changes restrict the forager's ability to recognize items, to do without nutrients, and so on. Other changed constraints reflect assumptions about the environment, for example, whether the forager finds prey in clumps or randomly dispersed. Constraints can be changed in many ways, and we can only discuss a few possibilities. We chose those "changed constraint" models that either make the basic models more realistic (they apply to many animals) or illustrate the assumptions and limitations of the basic models. We consider in section 3.2 sequential versus simultaneous encounter; in section 3.3 exclusivity of search and handling and the question of lost opportunity; in section 3.4 sequential dependence of encounters; in section 3.5 travel restrictions and central-place foraging; in section 3.6 nutrient and toxin constraints; and in section 3.7 recognition constraints. Chapter 4 deals with the problem of incomplete information.

3.2 Sequential versus Simultaneous Encounter

In prey and patch models, the forager encounters one item (a prey item or patch) after the other. Experiments can be devised to ensure that this assumption is met (e.g. Krebs et al. 1977), and sometimes sequential encounter is biologically appropriate (e.g. a shore bird probing for small invertebrates in the mud, Goss-Custard 1977a). However, many animals must experience simultaneous encounters. A honeybee foraging in a field of flowers may see several flowers at once (Waddington and Holden 1979, Stephens et al. 1986); a foraging bluegill sunfish may see more than one *Daphnia* at a time (O'Brien et al. 1976).

Many workers have proposed simultaneous encounter models (Waddington and Holden 1979, Carlsson 1983, Engen and Stenseth 1984, Stephens et al. 1986), and sometimes they have come to different conclusions. Waddington and Holden (1979) developed an optimality model to solve the problem illustrated in Figure 3.1. At the encounter point the forager sees two prey items: although $e_1/m_1 > e_2/m_2$ (m_i is the manipulation or "actual handling" time) the type 1 item is further away, so $e_1/(m_1 + \pi_1) < e_2/(m_2 + \pi_2)$ (π_i is the pursuit time.) In other words, at the moment of choice, the effective profitability of item 2 is higher than the effective profitability of item 1, because it is closer. Waddington and Holden claim that the general rule "choose the prey of higher effective profitability" is the rate-maximizing rule for simultaneous encounters.

However, Waddington and Holden's rule only holds at high encounter rates. When encounter rates are low, it is better to think of simultaneous encounters as a problem involving sequential encounters with sets of prey. Suppose that a raptor encounters flocks of small birds that contain two prey species; because an attack flushes the birds, the raptor may choose to attack an individual of species 1 or an individual of species 2, but not

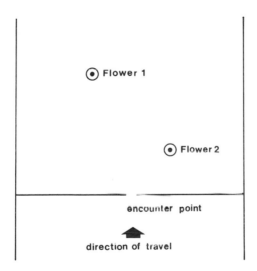

Figure 3.1 A foraging honeybee may simultaneously encounter a pair of flowers—Flower 2 which is closer and to the right and Flower 1 which is further away and to the left. According to Waddington and Holden's (1979) simultaneous encounter model, the more profitable flower should always be chosen. (Recall that profitability takes account of the time required to travel from the encounter point to the flower.) Thus Flower 2 may have a higher effective profitability simply because it is closer. (Redrawn from Stephens et al. 1986.)

both. Species 1 takes h_1 seconds to handle and yields e_1 calories, species 2 takes h_2 seconds and yields e_2 calories; e_1/h_1 is greater than e_2/h_2. We plot the two points (h_1, e_1) and (h_2, e_2) in Figure 3.2(A). The figure shows how the rate-maximizing choice of simultaneously encountered items can be found by constructing two lines, one from each of the points (h_1, e_1) and (h_2, e_2) to $(-1/\lambda, 0)$ ($1/\lambda$ is the time between encounters with flocks). The steepest line gives the rate-maximizing choice: if the steeper line passes

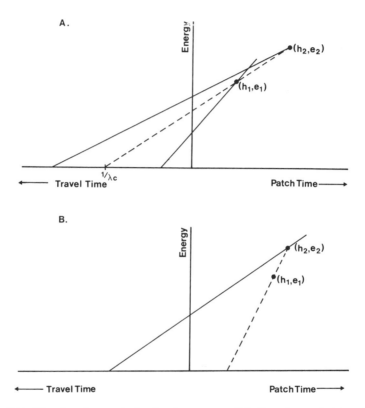

Figure 3.2 The discrete marginal-value theorem predicts the rate-maximizing choice between two simultaneously encountered and mutually exclusive choices. Each choice is characterized by an (h_i, e_i) pair, where $e_1/h_1 > e_2/h_2$, but $e_2 > e_1$. (A) The time intercept of the broken line passing through (h_1, e_1) and (h_2, e_2) defines a critical travel time $(1/\lambda_c)$. The solid lines show that travel times less than $1/\lambda_c$ yield a higher rate (the slope of the line) if type 1 is chosen, and that travel times greater than $1/\lambda_c$ yield a higher rate if type 2 is chosen. (B) A rate-maximizer cannot prefer a type that is both less profitable and has smaller energy value. The line of highest slope passes through (h_2, e_2), regardless of the travel time. As the dotted line shows, the travel time would have to be negative for the forager to prefer type 1!

through (h_1, e_1), then item 1 is the rate-maximizing choice; if the steeper line passes through (h_2, e_2), then item 2 is the rate-maximizing choice. If, as in Figure 3.2(A), e_2 is larger than e_1, even though type 1 is more profitable, then both the relative profitabilities of the two alternatives and the encounter rate with patches determine the rate-maximizing choice. Even though type 1 is more profitable, type 2 is the rate-maximizing choice at low encounter rates (Box 3.1, Orians and Pearson 1979, Engen and Stenseth 1984, Stephens et al. 1986).

Sequentially encountered sets of prey items are like patches, except that they can only be used in a limited number of ways. In the basic marginal-value theorem a forager can stay for any length of time from zero to infinity and obtain a continuous range of energy rewards, but our raptor can "stay" for only h_1 or h_2 and get only e_1 or e_2. In mathematical jargon, the gain function is discontinuous. In the previous paragraph and in Figure 3.2 we used a version of the marginal-value theorem for discontinuous gain functions to determine the rate-maximizing choice of prey items within a patch (Stephens et al. 1986, Fig. 3.2[A]). Engen and Stenseth (1984) have presented a general model that works for both continuous and discontinuous gain functions (Box 3.1). Discontinuous gain functions can arise for reasons other than simultaneous encounter with prey; they come up again in our discussion of central-place foragers (section 3.5). The marginal-value interpretation of simultaneous encounters predicts different patterns of preference under different conditions (Box 3.1). Waddington and Holden's "take the most profitable rule" will work at high encounter rates (with flocks), and it will work at any encounter rate if the most profitable choice also yields the largest net energy gain (Fig. 3.2[B]). In these cases, a forager only needs to "consider" immediate gains to maximize its long-term gains, but this is not always true (Box 3.1).

Simultaneous encounters can lead to an apparent violation of the zero-one rule, in the form of so-called partial preferences (Waddington 1982, Krebs and McCleery 1984). For example, a foraging honeybee may prefer blue flowers when they are paired with poor yellow flowers but ignore them when they are paired with rich purple flowers. Since the zero-one rule deals with the probability of attack upon encounter, and the pair, not the individual flower, is the encountered entity, the zero-one rule still holds: pairs of a given type are always attacked or always ignored. Partial preferences might also occur if the same pair is approached from different directions. If the pair in Figure 3.1 were approached from behind, it might be treated differently because the pursuit times would be different. Here, the zero-one rule still holds if we classify the "encountered entity" as "pairs attacked from a particular direction."

The preceding discussion has interesting implications for the most

BOX 3.1 THE DISCRETE MARGINAL-VALUE THEOREM

This box expands the model, discussed in the text, of a bird of prey attacking a flock. Suppose that a raptor encounters flocks consisting of two prey species (types): type 1 and type 2. Associated with each is a net energy value and a handling time: e_1, h_1, e_2, h_2. Type 1 is smaller but more profitable than type 2 ($e_2 > e_1$, but $e_1/h_1 > e_2/h_2$). Only one item can be attacked per encounter, because an attack flushes the flock: prey choices are *mutually exclusive*. Figure 3.2 shows the rate-maximizing prey choice.

EFFECT OF TRAVEL TIME
The time-intercept of the line passing through (h_1, e_1) and (h_2, e_2) defines a critical travel time (recall that travel time equals the reciprocal of encounter rate):

$$\frac{1}{\lambda_c} = \frac{e_1 h_2 - e_2 h_1}{e_2 - e_1}.$$

A rate-maximizer prefers the smaller, more profitable alternative when travel time is shorter than $1/\lambda_c$, and it prefers the larger, less profitable alternative when travel time is longer than $1/\lambda_c$.

ORDER OF ATTACK
Suppose that a honeybee encounters a pair of flowers like that in Figure 3.1, except that here the distant flower is both more rewarding (bigger e) and more profitable than the close flower. Since the bee can take both flowers, the flock model cannot be directly applied. Instead, we consider the four mutually exclusive alternatives: (A) take distant only; (B) take close only; (C) take close, then distant; and (D) take distant, then close. In general, this problem can be solved by plotting the pairs (e_A, h_A), (e_B, h_B), (e_C, h_C), and (e_D, h_D), as in Figure B3.1. A rate-maximizer should not even consider options (B) take close only and (D) take distant, then close, because the other options offer more energy and are more profitable. Thus the honeybee's choice becomes either (A) take distant only or (C) take close, then distant. Figure B3.1 shows the rate-maximizing behavior: Take (A) distant, when travel times are shorter than $1/\lambda_c$, and take (C) close, then distant, when travel times are longer than $1/\lambda_c$. Stephens et al. (1986) studied a honeybee flower choice problem like this one empirically.

The order of attack is sometimes used as a measure of preference: if an item is attacked first, it is said to be preferred. The example above shows that this makes no sense for a long-term rate maximizer. A long-term rate-maximizer may take the close, less profitable flower first, but it would never take the close, less profitable flower if the two flowers were mutually exclusive alternatives (Fig. 3.2[B]).

BOX 3.1 (CONT.)

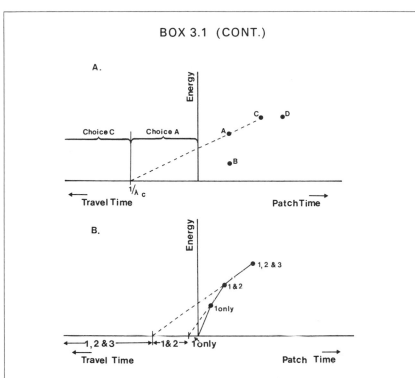

Figure B3.1 (A) Four (h_i, e_i) pairs are plotted for the four ways in which a pair of flowers (Figure 3.1) can be taken. Options B and D can never be the rate-maximizing choice. Option A is the rate-maximizing choice when travel times are shorter than $1/\lambda_c$. (B) If the order of attack does not affect the time spent in the patch, then the rate-maximizing behavior considers only which items to take. The slope of the line between adjacent points equals the profitability of the type just added.

In some experiments (e.g. operant simulations of foraging) the order of attack may not affect the time spent in the patch. Suppose that such a patch contains one item each of three different types: $e_1/h_1 > e_2/h_2 > e_3/h_3$. There are nine feasible orders of attack, but only three unique (h_i, e_i) pairs. Thus a rate-maximizer should consider only whether a type should be taken, and not when. The three options are

Tactic	h	e
1 only	h_1	e_1
1 & 2	$h_1 + h_2$	$e_1 + e_2$
1, 2, & 3	$h_1 + h_2 + h_3$	$e_1 + e_2 + e_3$

Figure B3.1(B) shows how patch encounter rates determine the rate-maximizing behavior. The slope of the line segment between adjacent points equals

---

BOX 3.1 (CONT.)

the profitability of the last item included. The order of attack does not matter to a rate-maximizer; it sets a profitability threshold: take all types that are more profitable than the long-term rate of energy intake.

ENGEN AND STENSETH'S CRITERION

Thus far we have only treated some special cases to illustrate how prey should be chosen within patches. Specifically, we have relied on "rooted-tangent" diagrams (section 2.3, Fig. 2.2), which are limited to the case in which all patches are the same, and in which between patch costs are nil.

Engen and Stenseth (1984) have proposed a general and elegant criterion for choosing the rate-maximizing patch-use tactic. Any possible patch-use tactic can be represented by a pair of numbers (h_i, e_i), where h_i is the handling or patch residence time for the ith patch tactic and e_i is the net energy gain for the ith patch tactic. Engen and Stenseth show that a long-term rate-maximizer will choose the pair (h_i, e_i) that maximizes

$$e_i - R^*h_i,$$

where R^* is the optimal long-term rate of energy intake. This term is the energy intercept of a line drawn through (h_i, e_i) with slope R^*. Engen and Stenseth's technique neatly subsumes both the discrete and continuous version of the marginal-value theorem, but it has the disadvantage of requiring that R^* be found (usually numerically) before it is of much use.

---

common type of laboratory experiment in the study of feeding behavior, the *dichotomous preference* test, because it predicts that preference sometimes depends on the schedule of presentation. If, for example, the experimenter allows the forager to consume one member of a pair and then waits a fixed time to present the next pair, the results of Figure 3.2 will apply. In other words, observed preferences may vary according to the inter-presentation interval. If, on the other hand, the experimenter allows a fixed time between presentations, then the type with the highest energy value should always be chosen as long as handling times are shorter than the interval between presentations, because no opportunity is lost.

Heller (1980) used computer simulations to study a similar problem. Heller's forager does not see all the prey when it encounters the patch (as we assume above); instead, it experiences strict sequential encounters within each patch. Heller considers a large patch that contains two prey types, and foraging in the patch depresses the within-patch encounter rates because capturing a type 1 prey item reduces the encounter rate with type 1's. Heller does not manipulate the encounter rate with patches; rather he considers changes in the relative abundance of the most profitable prey type within patches. Heller gives his forager four options: (1) specialize on the more profitable prey; (2) specialize on the less profitable prey; (3) generalize; and (4) specialize on the more profitable prey at first, but generalize later. Heller finds that the fourth tactic (switch from specialist to generalist) often does well in his simulations. Presumably this is because of depression in patch quality: the forager removes highly profitable prey first, thereby reducing their encounter rates; this leads to the inclusion of the lower-ranked prey. The interesting point here is that choosing prey within patches sometimes contradicts the zero-one rule. (Although at any one time it still applies, a Heller animal might go through cycles of "arrive patch 1: specialize-generalize," "arrive patch 2: specialize-generalize," and so on). Heller's simulation shows that this cyclical behavior can be the "long-term rate-maximizing" prediction, but it only hints at when it should occur.

3.3 Exclusivity of Search and Handling

Many foragers do not totally consume each prey item (Hassell et al. 1976), and some authors (Cook and Cockrell 1978, Giller 1980, Sih 1980) have used the marginal-value theorem to explain this. Foragers such as *Notonecta* and *Myrmeleon*, which suck the contents of their victims, can be shown by interruption experiments to experience diminishing returns with sucking time. This has led to the assumption that these predators treat their prey like patches. However, McNair (1983) and Lucas and Grafen (1985) have pointed out that these foragers differ from ordinary patch users in that ambush predators can sometimes encounter new patches (prey) while they handle an old one, and this violates the patch model's assumption that searching for and hunting within patches are mutually exclusive activities.

McNair supposes that foragers encounter new patches (prey items for a typical ambush predator) at rate λ after they have left a patch, as in conventional theory, but he also assumes that they can encounter patches at rate $\hat{\lambda}$ while they are exploiting a patch. When a *Notonecta* sucks the juices from its prey, it may still be able to encounter new prey items at the reduced rate of $\hat{\lambda}$; in McNair's terminology a second encounter can *overlap* the first.

Box 3.2 presents McNair's model of overlapping encounters. McNair's main conclusion is that the rate-maximizing forager should stay beyond the point at which the marginal capture rate (derivative of the gain function) equals the long-term average rate of energy intake. In other words,

BOX 3.2 OVERLAPPING ENCOUNTERS

McNair's (1983) overlapping encounter model might be applied to a forager like a web-building spider, which catches its prey in a trap. Since a spider eats by sucking the juices from its prey, and gains show diminishing returns with sucking time, a modified patch model can be applied to solve for the rate-maximizing prey-sucking time. Most spider webs work best when the spider is minding the lines (they are not completely passive traps), but some webs can capture prey even while the spider consumes a prey item, thereby violating the patch model's assumption that no new patches can be encountered during patch exploitation.

THE MODEL

Suppose, following the patch model, that the spider captures only one prey type, t is the sucking time (the decision variable), $g(t)$ is the gain function that relates sucking time to net energy gains, λ is the rate of prey capture when the spider is minding the lines, and $\hat{\lambda}$ is the reduced rate of prey capture when the spider is busy consuming a prey item. McNair compares the spider's capture process to humans queuing up in a fast-food restaurant: λ is the rate at which the first customer steps up to be served when the serving window is unoccupied, and $\hat{\lambda}$ is the rate at which customers join the queues behind busy windows. McNair thinks of each initial encounter as the beginning of an *extended encounter*. A result from the theory of queues (see McNair 1983) states that the spider can expect to encounter $\hat{\lambda}t/(1 - \hat{\lambda}t)$ additional prey items during an extended encounter (this result requires that $\hat{\lambda}t < 1$). Thus every time an initial encounter occurs the forager expects to capture $1 + [\hat{\lambda}t/(1 - \hat{\lambda}t)]$ prey items, or $1/(1 - \hat{\lambda}t)$. The long-term rate of energy intake is

$$R = \frac{\dfrac{\lambda g(t)}{1 - \hat{\lambda}t}}{1 + \dfrac{\lambda t}{1 - \hat{\lambda}t}} = \frac{\lambda g(t)}{1 + (\lambda - \hat{\lambda})t}.$$

Solving for the first-order maximization condition,

$$\frac{dR}{dt} = 0, \text{ where } g'(t) = \frac{g(t)}{\dfrac{1}{\lambda - \hat{\lambda}} + t}.$$

This expression suggests a simple change in the patch model's usual graphical solution. Figure B3.2 shows this solution (see section 2.3 for the limitations of this type of diagram). Instead of drawing a tangent to $g(t)$ from $1/\lambda$ on the travel time axis, the tangent is drawn from the "corrected travel time," $1/(\lambda - \hat{\lambda})$. The corrected travel time is always longer than the true travel time,

BOX 3.2 (CONT.)

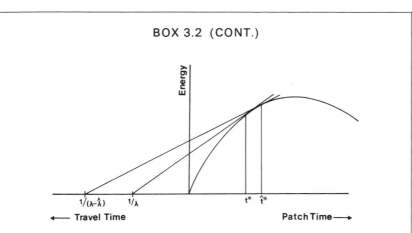

Figure B3.2 The rate-maximizing patch residence time for a forager experiencing overlapping patch encounters. The standard tangent solution is slightly modified (compare with Fig. 2.2). The tangent is drawn from the corrected travel time, $1/(\lambda - \hat{\lambda})$, which is always longer than the actual travel time. Therefore, the forager stays longer than it does in the conventional patch model.

so the model predicts that a rate-maximizer experiencing overlapping encounters should stay longer than one that does not. In the most extreme case, when $\lambda = \hat{\lambda}$, the forager should exploit the patch until the patch's energy reaches a peak $[g'(t) = 0]$. These predictions follow from the fact that a forager experiencing overlapping encounters loses less opportunity by staying in a patch even though returns are diminished.

a forager experiencing overlapping encounters should stay longer than predicted for the conventional marginal-value forager: it tolerates more patch depression.

This result seems to contradict the marginal-value theorem, because adding overlapping encounters ought to increase the long-term rate of energy intake, and an increased rate of energy intake ought to cause a decrease in patch residence time. However, McNair's result can be viewed as an exception that proves the rule. The marginal-value theorem works in the first place (i.e. a rate-maximizing forager leaves patches before they are empty) because by staying longer a forager loses the opportunity to find new patches. When a forager experiences overlapping encounters, it loses less opportunity by staying in patches, precisely because more patches can be encountered while the forager exploits a patch. Box 3.2 discusses additional "overlapping encounter" results.

Unfortunately, McNair's model assumes that the forager can capture new prey without losing its old prey. Lucas and Grafen (1985) have developed a model that assumes that the forager loses its present prey if it attempts to capture another. Lucas and Grafen assume that a forager will not switch to a new prey until it has exploited the old prey for some "criterion time," and they solve for the rate-maximizing criterion time. Their results agree qualitatively with McNair's: the rate-maximizing criterion time increases as the frequency of overlapping encounters $(\hat{\lambda})$ increases.

Lucas (1983) has analyzed another "less lost opportunity problem." Lucas imagines that a rate-maximizing forager feeds in short bouts, and he points out that the fussiness of a rate-maximizer should depend on bout length, because when there remains in a bout enough time only for one item to be taken, then no opportunity can by lost by taking a less profitable prey. This point is especially important for foragers that take only one item per bout, but Lucas shows that it can be generalized to more complex situations.

Pursuit and search. In Chapter 2 we included pursuit in our account of handling time. The basic prey model requires this, because pursuit clearly occurs only after the decision to attack has been made. However, pursuit differs from "handling" proper, because new items can often be encountered during pursuit. This difference might be important for animals that "pursue" stationary prey: a honeybee flying toward a flower it sees from 0.5 meters away may see a better flower before it reaches its intended victim. As in McNair's and Lucas's models of lost opportunity, pursuit involves less commitment and less lost opportunity than does manipulation or active chasing of a prey item. Although we know of no model that deals specifically with this issue, we expect that the ability to encounter items when pursuing others will tend to make rate-maximizing foragers less fussy about the items they pursue.

3.4 Prey Choice with Sequential Dependence

The basic prey and patch models assume that searching is like hunting through a randomly mixed sand pile grain by grain: most of the grains are undesirable (search), but occasionally the forager finds a desirable grain (i.e. a prey item or patch). The desirable grains are further subdivided into different types (e.g. red and blue "grains"). This means that prey (or patches) are well mixed in the habitat: the forager does not find a given type in clumps. McNair (1979) studied this problem by imagining that encounter rates might depend on the previous encounter. McNair imagines that the average times between (1) leaving a red and encountering a blue, (2) leaving a red and encountering a red, (3) leaving a blue and encountering a blue, and (4) leaving a blue and encountering a red may all be different, while in the basic "sand pile encounter" models (1) and (3) are always the same, as are (2) and (4), because encounter rates are independent of previous encounters. The encounter-to-encounter dependence assumed in McNair's model may occur because of prey clumping or over-dispersion. In Box 3.3 we present a two-prey-type version of McNair's more general model.

BOX 3.3 ENCOUNTER-TO-ENCOUNTER DEPENDENCIES

We illustrate McNair's (1979) general model by changing the prey model's usual Poisson encounter assumption and by a counter-example.

THE TWO-PREY-TYPE CASE

Suppose that there are two prey types, type 1 and type 2; as usual, $e_1/h_1 > e_2/h_2$. We still assume that a Poisson process controls encounters, but the encounter rate with each type depends on the most recently encountered type: leaving a type 1 item "starts the clock" for two independent Poisson encounter processes. The time, after a type 1, until another type 1 is encountered is exponentially distributed with rate parameter λ_{11}, and the time, after a type 1, until a type 2 is encountered is exponentially distributed with rate parameter λ_{12}. The notation λ_{ij} denotes the encounter rate for an i to j transition. This two-prey system's encounter process can be thought of as a matrix of encounter rates:

Encounter Rates (λ_{ij})

		To:	
		Type 1	Type 2
From:	Type 1	λ_{11}	λ_{12}
	Type 2	λ_{21}	λ_{22}

We want to calculate the probability of each type of transition (the P_{ij}'s). A transition from 1 to 2 occurs only if a type 2 is encountered after a type 1 but before the next type 1. If we let t_{ij} be a realization of the randomly chosen (and exponentially distributed) time from a type i encounter to a type j encounter, then the transition probabilities are: $P_{11} = P(t_{11} < t_{12})$, $P_{12} = P(t_{12} < t_{11})$,

BOX 3.3 (CONT.)

$P_{21} = P(t_{21} < t_{22})$, and $P_{22} = P(t_{22} < t_{21})$. We calculate only one value for illustration:

$$P_{12} = P(t_{12} < t_{11}) = \int_0^\infty (1 - e^{-\lambda_{12}t_{11}})\lambda_{11}e^{-\lambda_{11}t_{11}}\, dt_{11}$$

$$= 1 - \lambda_{11} \int_0^\infty e^{-(\lambda_{12}+\lambda_{11})t_{11}}\, dt_{11}$$

$$= \frac{\lambda_{12}}{\lambda_{11} + \lambda_{12}}.$$

The other four transition probabilities can be found in the same way:

Transition Probabilities (P_{ij})

		To:	
		Type 1	Type 2
From:	Type 1	$\dfrac{\lambda_{11}}{\lambda_{11} + \lambda_{12}}$	$\dfrac{\lambda_{12}}{\lambda_{11} + \lambda_{12}}$
	Type 2	$\dfrac{\lambda_{21}}{\lambda_{21} + \lambda_{22}}$	$\dfrac{\lambda_{22}}{\lambda_{21} + \lambda_{22}}$

This matrix of transition probabilities represents a first-order Markov Process. Well-known formulae give the stationary probabilities of a two-by-two Markov process (Cox 1962). The stationary, or equilibrium, probability of type 1's is the long-run relative frequency of type 1 encounters. Let π_1 and π_2 be the respective stationary probabilities:

$$\pi_1 = \frac{P_{21}}{P_{12} + P_{21}}$$

$$\pi_2 = \frac{P_{12}}{P_{12} + P_{21}}.$$

Finally, we must find the mean time from leaving an i until encountering a j. The logic here is like that used to find the P_{ij}'s; for example, we want to find the distribution of t_{12}, given that t_{12} is shorter than t_{11}:

$$P(t_{12}|t_{12} < t_{11}) = \frac{P(t_{12} < t_{11}|t_{12})P(t_{12})}{P(t_{12} < t_{11})}.$$

The denominator is simply the transition probability (P_{12}) that we found above. We can also find the numerator. The probability that t_{11} is greater than

BOX 3.3 (CONT.)

t_{12} is

$$1 - (1 - e^{-\lambda_{11}t_{12}}) = e^{-\lambda_{11}t_{12}},$$

and $P(t_{12})$ is just $\lambda_{12}e^{-\lambda_{12}t_{12}}$; thus

$$P(t_{12}|t_{12} < t_{11}) = \frac{\lambda_{12}e^{-(\lambda_{11}+\lambda_{12})t_{12}}}{\dfrac{\lambda_{12}}{\lambda_{12}+\lambda_{11}}}$$

$$= (\lambda_{12} + \lambda_{11})e^{-(\lambda_{11}+\lambda_{12})t_{12}}.$$

This expression shows that the search time of 1-to-2 encounters is exponentially distributed with parameter $(\lambda_{12} + \lambda_{11})$, and the mean search time is therefore $1/(\lambda_{12} + \lambda_{11})$:

Mean Search Times (S_{ij})

To:

		Type 1	Type 2
	Type 1	$\dfrac{1}{\lambda_{11}+\lambda_{12}}$	$\dfrac{1}{\lambda_{11}+\lambda_{12}}$
From:			
	Type 2	$\dfrac{1}{\lambda_{21}+\lambda_{22}}$	$\dfrac{1}{\lambda_{21}+\lambda_{22}}$

We now have enough information to specify the long-term rate of energy intake. The average energy gain per encounter is $\pi_1 e_1 + \pi_2 e_2$, and the average time per encounter is

$$\pi_1[P_{11}(S_{11} + h_1) + P_{12}(S_{12} + h_2)] + \pi_2[P_{21}(S_{21} + h_1) + P_{22}(S_{22} + h_2)].$$

Thus R is

$$\frac{\pi_1 e_1 + \pi_2 e_2}{\pi_1[P_{11}(S_{11} + h_1) + P_{12}(S_{12} + h_2)] + \pi_2[P_{21}(S_{21} + h_1) + P_{22}(S_{22} + h_2)]}.$$

Substituting these expressions, and performing some algebraic masochism, shows that the long-term average rate of energy intake for a diet of both types is

$$R = \frac{\phi_1(\lambda_{11} + \lambda_{12})e_1 + \phi_2(\lambda_{21} + \lambda_{22})e_2}{1 + \phi_1(\lambda_{11} + \lambda_{12})h_1 + \phi_2(\lambda_{21} + \lambda_{22})h_2},$$

where

$$\phi_1 = \frac{\lambda_{21}}{\lambda_{12} + \lambda_{21}} \quad \text{and} \quad \phi_2 = \frac{\lambda_{12}}{\lambda_{12} + \lambda_{21}}.$$

BOX 3.3 (CONT.)

But for a diet with only type i,

$$R = \frac{\lambda_{ii}e_i}{1 + \lambda_{ii}h_i}.$$

McNair's Counter-Example

Suppose that the matrix of encounter rates is

Encounter Rates (λ_{ij})

		To:	
		Type 1	Type 2
From:	Type 1	0.6	0.3
	Type 2	0.5	2.7

Notice that the rate of 2-to-2 transitions is more than four times faster than the next highest rate. Suppose also that $e_1 = 1$, $h_1 = 1$, $e_2 = 1$, and $h_2 = 1.1$. The three possible diets give long-term rates of gain of

	Diets:		
	1 only	2 only	Both
$R =$	0.3750	0.6801	0.6114.

This example violates the "ranking by profitability" result, because the least profitable type alone *can* be the rate-maximizing diet. It also violates the "inclusion is independent of encounter rate" prediction, since type 2's are included here because of their own high 2-to-2 encounter rates.

Box 3.3 shows two important conclusions. First, a type's inclusion in the diet generally depends on the encounter rates of *all* types: it no longer makes sense to talk about *the* encounter rate of a single prey type, (although equilibrium encounter rates for a given type can be specified), because a type's encounter rate depends on encounters with other types. In other words, the independence of inclusion from encounter rate (section 2.2) of the basic prey model fails in this model. Second, the ranking of prey by profitability also fails. A counter-example demonstrates this failure: the basic prey model says that a forager's diet cannot include the less profitable type 2 unless it also includes the more profitable type 1. In the example in Box 3.3 encounters between two type 2's (2 to 2 transitions) are fast, but encounters between other pairs (1 to 2, 1 to 1, 2 to 1) are slower. We can think of this in the following way: pairs of type 2's become the most favored prey "items," and the forager ignores profitable type 1 prey because taking a type 1 risks a missed opportunity to take a highly profitable pair of type 2's.

McNair's analysis models the problem of an animal that feeds on loosely defined clumps of prey rather than on the discrete and well-defined patches of conventional theory. However, a crucial feature of this model is that the encounter-to-encounter dependency is simply a statistical phenomenon. Encountering a blue tells the forager the probability distributions of the times until the next encounter with a blue and with a red. The model does not apply to clumps in which the forager can recognize all the members of the clump (the results of section 3.2 would apply). The model may apply to a forager that only recognizes prey items when it bumps into them, but even in this case the model sweeps some interesting biology under the rug. For example, suppose that prey always occurs in pairs: after finding an item, the forager might search carefully for the second member of the pair, and this may influence the distribution of search times.

The zero-one rule still holds in McNair's model, because each encounter with a blue gives the forager the same information about the expected time to the next red and blue. This is because McNair's model assumes only first-order dependence. A more general nth order dependence model would be complicated. Lucas (1983) has discussed some simple encounter patterns that illustrate the possible effects of "many encounter" dependencies. Consider the following string of encounters: red–1, red–2, red–3, blue–1, blue–2, blue–3, blue–4, red–4, red–5, red–6, blue–5, blue–6, blue–7, blue–8. The handling times are equal for reds and blues, but reds provide higher net energy (reds are more profitable). Handling takes just long enough so that at best only every other item can be taken. The rate-maximizing policy here violates the zero-one rule: Take red–1 (red–2 is missed because red–1 is being handled), take red–3 (blue–1 is missed), take blue–2 (blue–3 is missed), *ignore blue–4* because taking it would mean getting only one red from the next three. The point here is that if the forager has some special knowledge about the sequence of items, it might use this knowledge to determine whether anything is to be lost by taking a particular item (Lucas 1983).

3.5 Travel Restrictions and Central-Place Foraging

One interesting and general change in the basic models is related to the problem of travel restrictions. A parent bird must return time after time to its nest with food for its young; a waiting *Anolis* lizard may see prey from its perch on a tree trunk, attack it, and return to its perch. In cases like these the two basic models can be extended to make specific predictions about how patches should be used and at what distances from the *central place* items should be attacked. Rather than violating the basic assumptions, these constraints extend the basic models to a new class of problems. Three central-place models have been proposed. The first of these is Schoener's (1979) model of "encounter at a distance," which might be applied to the lizard example above. The second and third are Orians and Pearson's (1979) two central-place foraging models that deal with two different kinds of foragers, the single-prey loader and the multiple-prey loader.

ENCOUNTER AT A DISTANCE

In the basic prey model prey types can be characterized by their profitabilities (e/h). Consider a case in which prey can be recognized from varying distances. A cricket of a certain size encountered from a distance of 2 meters will have a different e/h than the same cricket seen from a distance of 2 centimeters. Distance at encounter affects both net energy gain (e) and handling time (h): pursuit from a greater distance will take longer and cost more. Schoener's model explores the implications of these distance-related changes.

Schoener initially assumes that the rate-maximizing rule is of the form "take all prey above some critical e/h value" (recall that h explicitly includes all the time from encounter until searching resumes). His model also assumes sequential encounters, and thus he excludes the simultaneous encounter problems considered in section 3.2. Suppose that a flycatcher sallies for insects from a fixed perch. Prey types, each with a characteristic length and encounter rate (prey can be categorized by their lengths, because length is closely related to energy value), are briefly accessible at the central point of a conveyor belt located d meters away from the perch. Schoener's model considers the consequences of moving the conveyor belt both closer and further away.

When certain relationships exist between prey length and handling time, between prey length and energy value, and between the relative costs in energy of manipulating and pursuing (Schoener discusses these relationships), then the relationship between prey length and e/h is unimodal, as Figure 3.3 shows. Both small and large prey are less profitable than intermediate sizes. Figure 3.3 also shows how the relationship between prey length profitability changes with the distance (d) between the flycatcher's perch and the conveyor belt: as the values of d increase, profitability generally decreases (the "far" curve is everywhere below the "close" curve), and the most profitable prey length becomes larger as d increases.

The flycatcher example can be generalized by imagining that there are two conveyor belts at different distances (close and far) in front of the fly-

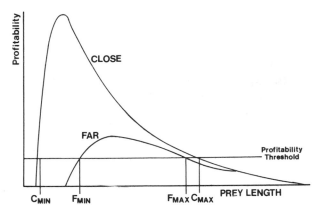

Figure 3.3 Hypothetical relationships between prey length and prey profitability. This relationship depends on the distance at encounter: an item encountered from far away is less profitable than the same item when close. C_{min}, C_{max} are the minimum and maximum acceptable prey lengths for "close." F_{min} and F_{max} are the minimum and maximum acceptable prey lengths for "far." Any length acceptable from far away is also acceptable at a closer distance.

catcher. Applying the threshold profitability rule of the basic prey model, we can use Figure 3.3 to find the acceptable prey lengths at each distance. A line drawn parallel to the prey length axis represents this threshold. The figure shows that the range of acceptable prey lengths is smaller at greater distances, the sizes of prey taken at greater distances are in general larger, and the range of acceptable lengths at a given distance is always wholly within the range of acceptable sizes for a shorter distance. Any item acceptable at 20 meters must be acceptable at 15 meters. Although in nature flycatchers seldom feed from conveyor belts, we use this example to emphasize the comparison between close and far prey. Schoener's model can easily be generalized to more natural situations.

Our discussion of Schoener's model treats what we consider the most general case, but there are other possibilities. For example, if pursuit costs increase with prey length, then Schoener's "size-distance" effect can be reversed—the forager accepts smaller prey at greater distances—but it still holds that any prey length acceptable at a great distance must be acceptable at a closer distance (see Schoener 1979).

THE SINGLE-PREY LOADER AND SIMULTANEOUS ENCOUNTER

Many birds feed their nestlings by traveling from the nest to patches of food and, after spending some time hunting in the patch, returning to the nest with a *single* prey item. Orians and Pearson (1979) called such animals "single-prey loaders." The single-prey loader's problem can be solved using the techniques we introduced to treat simultaneous encounters (section 3.2). Like a forager experiencing simultaneous encounters, the single-prey loader exploits patches with discontinuous gain functions; unlike a forager experiencing simultaneous encounters, the single-prey loader must search within the patch for a single prey item.

The effect of within-patch search. Orians and Pearson's single-prey loader problem imagines that the forager travels to some distant patch, searches there for prey items, and returns with a single item. In simultaneous encounter models the forager simply picks an item (in the simplest case) without having to search for it. Being unselective with sedentary, simultaneously encountered prey is never the rate-maximizing policy (except when $\lambda = \lambda_c$; see Fig. 3.4[A]), because the average energy gain ($\bar{e}$) and average handling time ($\bar{h}$) for an unselective policy (such as take type 1 or type 2 with equal probability) lie on the line in Figure 3.4(A). For the single-prey loader, however, total within-patch time (h_i for patch-use tactic i) is the sum of two components, within-patch search time ($1/\beta_i$) and prey manipulation time (m_i). When the forager must search within the patch, being unselective can reduce search time. For example, if the mean time to the first encounter with a type 1 ($1/\beta_1$) is 5 seconds and the mean time to the first encounter with a type 2 ($1/\beta_2$) is 10 seconds, then the expected time to the first encounter regardless of type $[1/(\beta_1 + \beta_2)]$ is 3.3 seconds, and not the average of these times (7.5 seconds). Thus ($\bar{h}, \bar{e}$) lies above the line between (h_1, e_1) and (h_2, e_2) (see Fig. 3.4).

This outcome means that the patch-use tactic "be unselective" will often be the rate-maximizing choice. The single-prey loader may be unselective when patches are close but select for large type 2's when patches are far away (Krebs and Avery 1985). On the other hand, it may select for small, profitable type 1's when patches are close, be unselective when patches are at intermediate distances, and select for large type 2's when patches are far away (Fig. 3.4[B]). This change from selectivity to unselectivity and back to selectivity can only occur when the manipulation time for type 2's (m_2) is greater than the manipulation time for type 1's (m_1). (This limitation arises because the tactic "attack only type 1's" is both less profitable and less rewarding than the tactic "be unselective" when $m_2 \leq m_1$: see section 3.2.) Despite the possibility of a change from selectivity to unselectivity and back to selectivity (Fig. 3.4[B]), the average prey size (energy value) taken generally increases with average distance from the central place. We can still conclude, then, that size-selectivity increases with distance.

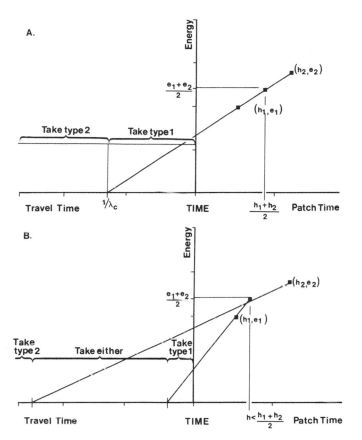

Figure 3.4 This diagram compares the "simultaneous encounter" and "single-prey loader" problems. (A) Simultaneous encounter: the expected "within-patch" time ($\bar{h}$) and energy gain ($\bar{e}$) [i.e. the pair ($\bar{h}$, $\bar{e}$)] lie on the line between (h_1, e_1) and (h_2, e_2), because the forager does not search within the patch. (B) Single-prey loader: when there is within-patch search, being unselective reduces the expected patch time below the mean within-patch time of the two selective policies. This means that at intermediate travel times unselective behavior can be predicted when $m_2 > m_1$ (m_i – loading or manipulation time for type i).

A simpler model for the single-prey loader has proven valuable for those cases in which manipulation times are the same for all prey types, and in which a forager may choose prey from a continuous range of prey sizes (Orians and Pearson 1979, Lessells and Stephens 1983). When manipulation times are all the same, it is always better to take larger prey, and since prey sizes are available from a continuous range, the model solves for the rate-maximizing "minimum acceptable prey size." Box 3.4

BOX 3.4 THE SINGLE-PREY LOADER

Suppose that a single-prey loader travels to a patch and searches within that patch for prey representing a broad range of energy values. For simplicity, suppose that loading a prey item takes no time. (A model similar to the one presented in this box can be built as long as loading times are constant for all prey sizes.) Thus searching occupies all the forager's patch residence time. The forager can use energy values to rank prey, and this ranking suggests a

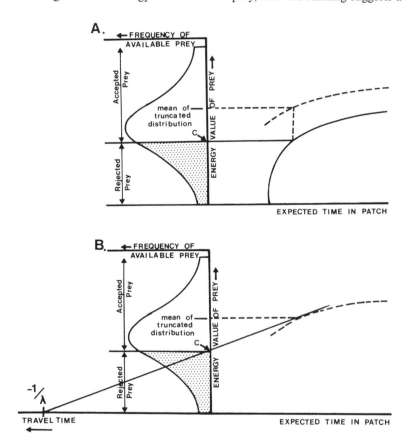

Figure B3.4   (A) The solid curve shows the relationship between the minimum acceptable energy value (C) and the mean patch residence time. The dashed curve plots the mean patch residence time against the mean energy value. The mean energy value is the mean of the prey energy value distribution truncated below C. (B) The standard tangent solution. Notice that the tangent intercepts the energy axis at C.

BOX 3.4 (CONT.)

simple patch-leaving rule: "Search until an item containing C or more calories is captured." The negligible-handling-time/single-prey loader model solves for a rate-maximizing "minimum acceptable energy value" (C^*).

The minimum acceptable energy value and the frequency distribution of energy values determine both the mean energy gain from a patch visit ($\bar{e}$) and the mean within-patch search time ($\bar{t}$). The mean energy value ($\bar{e}$) is the mean of the prey energy value distribution truncated below C, and the minimum acceptable energy value affects the within-patch search time ($\bar{t}$) because it determines the proportion of prey to ignore (the cumulative distribution function of the energy value distribution). A plot of ($\bar{t}$, $\bar{e}$) pairs generally looks like the dashed curve in Figure B3.4(A), and the solid curve gives the relationship between the decision variable C and the mean patch residence time $\bar{t}$. Figure B3.4(A) also shows how the energy value distribution affects these relationships.

Figure B3.4(B) shows the usual "tangent solution" to find the rate-maximizing ($\bar{t}$, $\bar{e}$) pair. Once this pair has been found, the C-curve in Figure B3.4(A) yields the rate-maximizing minimum acceptable energy value (C^*). However, C^* can be found more easily, because the energy-intercept of the tangent equals C^* (see Lessells and Stephens 1983 for proof).

This graphical solution gives two results: (1) the minimum acceptable prey size increases with travel time and (2) a rate-maximizing forager should be unselective below some critical travel time (T_{crit} in Fig. B3.4[B]). See Lessells and Stephens (1983) for a complete discussion of this solution.

develops this model, showing that a rate-maximizing forager ought to be unselective below some critical travel time (T_{crit}), and that the rate-maximizing "minimum acceptable prey size" generally increases with travel time.

Both travel restriction models discussed so far reach similar (but not identical) conclusions. Orians and Pearson (1979) express a general principle of rate-maximizing prey models: "For short travel times, superiority of prey hinges on energy per unit (true) handling time. For long travel times, superior prey are those of higher energy, regardless of handling time."

Encounter at a distance versus the single-prey loader. Orians and Pearson's single-prey loader model has been more widely tested than Schoener's encounter at a distance model. This is partly because the single-prey loader model applies directly to nesting birds. Although Schoener has suggested that the two models might be treated as theoretical alternatives, they make different decision assumptions. Schoener's model asks whether items should be attacked or ignored *when encountered* at varying pursuit distances; it is the prey decision made from a central place. The Orians and Pearson model asks how patches, themselves at given distances from a central place, ought to be exploited. For example, a forager might face an "Orians and Pearson" decision nested within a "Schoener" decision. Imagine that, in addition to a larger set of prey items, our flycatcher's conveyor belt occasionally serves up some pairs of prey items. Schoener's model tells us whether the pair should be attacked or ignored. Orians and Pearson's model tells us which member of the pair should be taken.

THE MULTIPLE-PREY LOADER

Some parent birds, as well as other central-place foragers, can carry many items at once: swallows collect insect prey for their young in a large bolus; starlings line many mealworms up in their beaks; chipmunks stuff sunflower seeds in their cheek pouches to take back to their larders. These animals are called multiple-prey loaders.

Orians and Pearson (1979) suggest that a loading curve characterizes multiple-prey loading. The loading curve, like the patch model's gain function, is assumed to be negatively accelerated, but for a different reason. The marginal loading rate decreases because the load itself hinders further loading. For example, a beak loaded with prey reduces a starling's ability to find and collect soil invertebrates, because the bird cannot readily probe (*Zirkeln*, Lorenz 1949), and this effect is cumulative with load size (Tinbergen 1981). For multiple-prey loaders, patch residence time determines load size, and the marginal-value theorem can be applied (section 2.3). The chief prediction is that increasing average distance from the central place should be matched by increasing load size.

In central-place foraging models it is tempting to predict, for example, that "foragers should stay longer in distant patches," but this is not strictly true. Suppose that a multiple-prey loader alternates between trips to a distant patch and trips to a close patch, and that these patches have the same loading curves. Should the forager take bigger loads from the distant patch? The answer is no, because the leaving rule for a patch will be based on the long-term average rate of energy intake, and this rate must be the same for both patches (since it is defined for the habitat that contains them). The predictions about the relationship between load size and distance apply to the *average* patch distance in the habitat, and not to individual trips (Lima 1982, personal communication).

3.6 Nutrients and Toxins as Constraints

The rate-maximizing models measure benefit in units of energy, but requirements for particular nutrients or the avoidance of toxins may also affect a forager's diet. One way to model this problem is to change the currency to allow the value of foraging tactics to reflect factors other than time and energy (Chapters 5 and 6). Another is to assume that nutrients simply limit the feasible diets that a rate-maximizer may choose. According to this view, if a particular prey type does not yield enough of the limiting nutrient it is not considered a feasible alternative.

Pulliam (1975) presented a "nutrients as constraints" model, which is shown in Box 3.5. Only one prediction differs from the basic prey model: there may be a partial preference (a violation of the zero-one rule) for the less profitable prey item if it is an important source of the limiting nutrient(s). The forager takes the more profitable item on every encounter, but a nutrient requirement means that the forager must sometimes take the less profitable type to compensate for the profitable type's low nutritional value. It is not surprising that nutrient requirements do not result in the more profitable but less nutritious types being taken on fewer than 100% of encounters: in the sequential encounter prey model the two types are not mutually exclusive alternatives (although taking more type 2's

---

BOX 3.5 NUTRIENT CONSTRAINTS

Chapter 2 proved the prey model's zero-one rule by showing that the sign of $\partial R/\partial p_i$ (R = long-term rate of energy intake and p_i = the probability of attacking a type i upon encounter) is independent of the magnitude of p_i. This means that either the maximum p_i or the minimum p_i should be chosen by a rate-maximizer. The zero-one rule is a special case of a min-max rule: if nothing else restricts p_i, then it should be set either to zero or to one.

NUTRIENTS

Pulliam (1975) argued that nutrient requirements often may restrict the possible values of p_i. Consider the two-prey-type case:

$$\text{If } e_1/h_1 > e_2/h_2, \text{ then } p_1^* = p_{1\ max},$$

$$\text{and (i) if } \frac{p_{1\ max}\lambda_1 e_1}{1 + p_{1\ max}\lambda_1 h_1} < \frac{e_2}{h_2}, \text{ then } p_2^* = p_{2\ max},$$

$$\text{and (ii) if } \frac{p_{1\ max}\lambda_1 e_1}{1 + p_{1\ max}\lambda_1 h_1} > \frac{e_2}{h_2}, \text{ then } p_2^* = p_{2\ min}.$$

BOX 3.5 (CONT.)

Figure B3.5(A) illustrates the usual situation. The shaded square represents all the feasible (p_1, p_2) choices. Here only the nature of probabilities $(0 \leq p_i \leq 1)$ limits the p_i's, and the min-max rule becomes the zero-one rule.

Now suppose that the forager needs at least N units of nutrient per unit of *search* time: then p_1 and p_2 must satisfy

$$p_1 \lambda_1 n_1 + p_2 \lambda_2 n_2 > N,$$

where n_1 is the amount of nutrient in type 1 items and n_2 is the amount of nutrient in type 2 items. Figure B3.5(B) illustrates this situation. The shaded area contains all those (p_1, p_2) pairs satisfying the inequality above. Notice that $p_{1 \text{ max}}$ and $p_{2 \text{ max}}$ are not affected (they still equal 1). Thus the predictions are changed only if $p_{2 \text{ min}}$ is the rate-maximizing choice [case (ii) above]. If type 2 items are an important source of the required nutrient, then the constrained rate-maximizer should attack them, even if attacking them decreases the rate of energy intake.

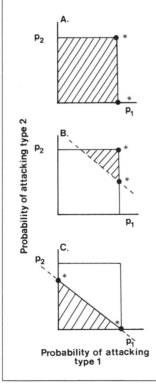

Figure B3.5 The shaded areas represent the feasible choices of (p_1, p_2) pairs (p_i is the probability of attacking type i upon encounter). Asterisks show the possible rate-maximizing solutions. (A) The conventional prey model: $0 \leq p_i \leq 1$, $p_1^* = 1$, p_2^* may be 1 or 0. (B) Nutrient constraints: the shaded area represents diets that provide enough nutrient. The solution is $p_1^* = 1$, but p_2^* may be 1 or $p_{2 \text{ min}}$. (C) Toxin constraints: the shaded area represents diets that keep the intake of toxin at or below a tolerable level. The solutions are $p_1^* = 0$ and $p_2^* = p_{2 \text{ max}}$, or $p_1^* = p_{1 \text{ max}}$ and $p_2^* = 0$.

BOX 3.5 (CONT.)

TOXINS

Suppose that the forager can tolerate only N units of some toxin. Here we reverse the direction of the constraint inequality. Figure B3.5(C) shows the feasible (p_1, p_2) pairs for the toxin tolerances: the feasible solutions now lie below the constraint line. If $p_{2\,min}$ is the rate-maximizing choice [case (ii) above], then the shaded triangle's lower right-hand vertex is the solution. However, if $p_{2\,max}$ is the rate-maximizing choice [case (i) above], then the solution is not obvious because $p_{2\,max}$ and $p_{1\,max}$ cannot be chosen simultaneously: if $p_{1\,max}$ is chosen, then p_2 must be set to zero. Because of the toxin constraint, the forager must reduce p_1 to increase p_2. Is this worthwhile? The solution must lie on the constraint boundary, because at a given p_1 the rate-maximizing p_2 is the maximum feasible p_2, and this always lies on the boundary line. We substitute the equation of the constraint line into Holling's disc equation, and this expresses the long-term rate of gain in the single variable p_1. Differentiating with respect to p_1 shows that the sign of $\partial R/\partial p_1$ is independent of the magnitude of p_1, so $p_{1\,max}$ and $p_{1\,min}$, marked by asterisks in Figure B3.5(C), are the only possible solutions. Some algebra shows that $p_{1\,max}$ should be chosen if

$$n_1 e_2 - n_2 e_1 < N(e_1 h_2 - e_2 h_1).$$

The right side of this expression is always positive (since $e_1/h_1 > e_2/h_2$); thus if type 1 items have more calories per unit of toxin than type 2 items $(e_1/n_1 > e_2/n_2)$, then $p_{1\,max}$ is the rate-maximizing solution. However, if type 2 items have more calories per unit of toxin, then type 2 can be the rate-maximizing choice to the exclusion of type 1. Along the constraint boundary, neither encounter rate has any effect on which type is the rate-maximizing choice, although the encounter rates determine the slope and intercepts of the boundary line.

does mean taking fewer type 1's), and this model ignores the temporal pattern of nutrient acquisition. For example, the forager may need m units of sodium per day, but it does matter whether the forager obtains this sodium in one lump or in a smooth flow throughout the day.

In Pulliam's model, nutrients differ from toxins only in that they constrain the forager from a different direction. A *nutrient* is a substance of which an animal must have at least a fixed amount, but a *toxin* is a substance of which an animal can tolerate at most a fixed amount. The toxin results differ from the nutrient results, because toxins can cause partial preferences for both prey types (Box 3.5).

3.7 Recognition Constraints

An important category of "changed constraint" models places limits on prey recognition. These models bring foraging theory to bear on the phenomenon of cryptic prey, and they relate foraging theory to the sensory limitations of foragers. There are two groups of recognition constraint models: (1) recognition time models (Charnov and Orians 1973, Elner and Hughes 1978, Hughes 1979, Erichsen et al. 1980, Houston et al. 1980) and (2) imperfect resemblance models (Getty and Krebs 1985, Getty 1985).

NON-ZERO RECOGNITION TIME

Figure 3.5 shows the sequence of choices made by a forager attacking a prey item (Charnov and Orians 1973). The basic prey model (Chapter 2) uses the first branching point in this hierarchy as its decision variable (p_i = the probability of initiating an attack), and it assumes that the expected "involvement time" (Hughes 1979) due to an encounter is $p_i h_i$ (h_i—the handling time—includes pursuit, attack, and consumption times). This assumption requires that the recognition time (r) be zero. If recognition time is not zero, then the expected involvement time becomes $r + p_i h_i$; since r is not multiplied by p_i, a non-zero recognition time limits the forager's control over the rate of energy gain: the forager pays the cost of recognition time regardless of its decision.

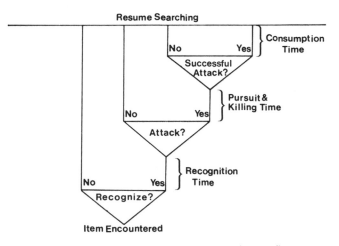

Figure 3.5 A simple hierarchy of "attack." The forager first recognizes an item as prey, then decides whether to attack and pursue it, and then, if the attack is successful, consumes it. The basic prey model assumes that recognition is instantaneous.

Many foragers must spend some time recognizing their prey. Shore crabs (*Carcinus maenus*) can tell good mussels from bad mussels by lifting them, because heavier mussels are more profitable, but this lifting takes time (Elner and Hughes 1978). Erichsen et al. (1980) and Houston et al. (1980) manipulated recognition time experimentally by presenting great tits (*Parus major*) with big or small pieces of mealworm hidden inside either opaque or clear drinking straws. The great tits had to pay the cost of recognition time when they encountered opaque straws.

A non-zero recognition time violates the prey model's prediction that "inclusion is independent of encounter rate." To demonstrate this, we follow Houston et al.'s (1980) development. A rate-maximizing forager should specialize on type 1 if

$$\frac{\lambda_1 e_1}{1 + \lambda_1(r + h_1) + \lambda_2 r} > \frac{\lambda_1 e_1 + \lambda_2 e_2}{1 + \lambda_1(r + h_1) + \lambda_2(r + h_2)}. \tag{3.1}$$

We only consider the case in which recognizing a type 1 item takes the same time as recognizing a type 2 item (see Houston et al. 1980 for other cases). Some algebra shows that expression (3.1) can be rearranged as

$$\lambda_1 > A + Ar\lambda_2, \tag{3.2a}$$

where

$$A = \frac{1}{\dfrac{e_1 h_2}{e_2} - (h_1 + r)}. \tag{3.2b}$$

The straight line specified by expression (3.2) divides those (λ_2, λ_1) pairs implying specialization (above the line) from the (λ_2, λ_1) pairs implying generalization (below the line; see Fig. 3.6). When recognition time is nil ($r = 0$), this model becomes the basic prey model, and the "independence of inclusion from encounter" prediction is preserved, because the specialization-generalization boundary is parallel to the λ_2 axis when $r = 0$. When recognition time is not zero ($r > 0$), the encounter rate with type 2 items (λ_2) affects whether type 2's are attacked. Figure 3.6(B) shows this effect: at a fixed value of λ_1 (say, $\hat{\lambda}_1$), type 2's will be attacked at high encounter rates ($\lambda_2 > \hat{\lambda}_2$) and ignored at low encounter rates ($\lambda_2 < \hat{\lambda}_2$). This happens because the time cost of rejecting a type 2 is too large to make rejection worthwhile. Moreover, the forager is unlikely to specialize at high recognition times, because specialization requires higher type 1 encounter rates when recognition times are high.

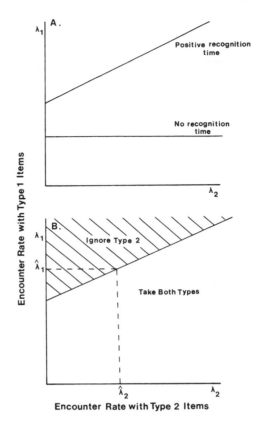

Figure 3.6 The lines divide the pairs of encounter rates (λ_2, λ_1) that predict special-ization (above the line) from the pairs that predict generalization (below the line). (A) When recognition is instantaneous, the encounter rate with type 2 items (λ_2) does not affect the inclusion of type 2, since the line is parallel to the λ_2 axis. When there is a positive recognition time, the diet depends on the encounter rates with both types. (B) The lower panel shows how the encounter rate with type 2 items can affect their inclusion in the diet. When the encounter rate with type 2 items exceeds $\hat{\lambda}_2$ and the encounter rate with type 1 items equals $\hat{\lambda}_1$, type 2's are attacked when encountered.

IMPERFECT RESEMBLANCE

The anthropologist J. Martin (1983) criticized the prey model because he thought it predicted that a foraging Eskimo must ignore a beached whale if swimming whales are not in the rate-maximizing diet. According to the basic models (Chapter 2), the forager's sensory abilities define types, and Martin, in effect, assumes that Eskimos cannot tell the difference between a beached and a swimming whale: prey types must be categorized by the forager's sensory capabilities. Types need not be biological species. Although Martin's mistake is obvious, there are subtle and interesting questions about how *types* should be defined (Getty and Krebs 1985, Getty 1985).

Getty argues that we might define types in one of two ways: a type can be defined by its appearance (or other sensory information), as we have done (Chapter 2), or by its profitability—for example, all potential prey with a profitability of 2.34 calories/second would be the same type regardless of what they looked like (Bell et al. 1984). Appearance types might be related to profitability types in three ways (Fig. 3.7). First, there may be a simple one-to-one relationship between appearance and profitability: everything that has appearance $a1$ has profitability $p1$, and everything that has appearance $a2$ has profitability $p2$ (perfect resemblance). Second, appearance may not be informative: prey are equally likely to have profitability $p1$ or $p2$ regardless of their appearance (Getty calls this "perfect

Figure 3.7 Hypothetical relationships between appearance types and profitability types. The shaded area represents (for example) the proportion of appearance types $a1$ that are of profitability type $p1$. (A) Perfect resemblance: all $a1$ appearance types have profitability $p1$. (B) Perfect mimicry: appearance gives no information about profitability. (C) Imperfect resemblance: an $a1$ may have profitability $p1$ or $p2$, and the probability of profitability $p1$, given appearance $a1$, is generally different from the probability of profitability $p1$, given appearance $a2$. Thus, appearance is partially informative.

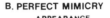

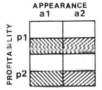

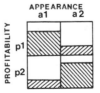

mimicry"); students of mimicry obviously prefer to think of profitability types. Third, appearance may give partial information about profitability: of the items with appearance $a1$, 70% have profitability $p1$ and 30% have profitability $p2$; of the items with appearance $a2$, 10% have profitability $p1$ and 90% have profitability $p2$ (imperfect resemblance). Should we use appearance or profitability types? Getty shows that, although we may often be interested in profitability types for other reasons, only appearance (or sensory) types preserve the zero-one rule of the basic prey model.

Suppose that a forager eats green caterpillars, but some green caterpillars are profitable and others are unprofitable. Caterpillar greenness ranges from pale green to dark green, but, although pale green caterpillars are usually more profitable, greenness gives imperfect information about profitability because the correlation is sloppy. Figure 3.8(A) shows two hypothetical overlapping "greenness" distributions, one corresponding to profitable types and the other corresponding to unprofitable types. Now suppose that the forager sets a greenness threshold: it eats caterpillars that are paler than the threshold and ignores caterpillars that are darker than the threshold. The threshold's position affects two crucial variables, the probability of correctly attacking a good item when encountered [which Getty calls P(Hit) following the terminology of signal detection theory—Egan 1975] and the probability of incorrectly attacking a bad item when encountered [called P(False Alarm)].

Following signal detection theory, the relationship between P(False Alarm) and P(Hit) characterizes a given discrimination problem. Figure 3.8(B)–(E) shows how this relationship follows from the greenness threshold discussed above. Signal detection theorists (Egan 1975) call the curve that relates P(False Alarm) to P(Hit) the "receiver operating characteristic," or ROC curve. A power law, $P(\text{Hit}) = [P(\text{False Alarm})]^k$, represents this relationship in a simple and convenient way, with k measuring the "discriminability" of the system. When k equals 1, hits and false alarms are equally likely (perfect mimicry—the ROC curve is a straight line with a slope of 1 that passes through the origin). When k equals 0, P(Hit) is always 1 regardless of the frequency of false alarms (perfect resemblance). However, when k is between 0 and 1, the discrimination is partial (imperfect resemblance). The more bowed-out the ROC curve, the more discriminable the prey.

To summarize the green caterpillar example, there are two distinct profitability types that overlap in appearance, the forager has a rule (the greenness threshold) that allows partial discrimination of the profitability types, a particular greenness threshold specifies a point on the ROC curve, and changes in the greenness threshold trace out the ROC curve. In turn, the ROC curve says something about the discriminability of the system.

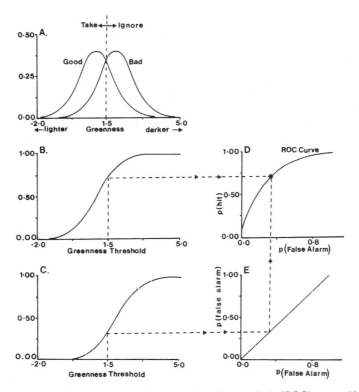

Figure 3.8 The derivation of a receiver operating characteristic (ROC) curve. (A) Two overlapping greenness distributions. Light green caterpillars (the distribution labeled "good") are more profitable than dark green caterpillars (the distribution labeled "bad"). (B and C) The forager discriminates by setting a greenness threshold: caterpillars paler than the threshold are eaten. (B) shows the relationship between the greenness threshold and the probability of eating a "good" upon encounter, P(Hit). Curve B is the cumulative distribution function of the "good" distribution in (A). (C) shows how the greenness threshold affects the probability of eating a "bad" upon encounter, P(False Alarm). (D and E) show how a given greenness threshold determines a point on the ROC curve. The dashed lines from (B) to (D) show how a threshold of 1.5 determines P(Hit). The dashed lines from (C) to (E) and then to (D) show how the threshold determines P(False Alarm). (E) is a reflection line that translates P(False Alarm) from the ordinate to the abscissa of (D).

How should a rate-maximizer set the greenness threshold? Finding the rate-maximizing threshold is equivalent to finding the rate-maximizing P(False Alarm), because the rate-maximizing P(False Alarm) can be used to find both P(Hit) and the rate-maximizing greenness threshold. The rate-maximizing P(False Alarm) can be found using an extension of the marginal-value theorem. A given value of P(False Alarm) sets both the

BOX 3.6   RATE-MAXIMIZING DISCRIMINATION STRATEGIES

This box considers the rate-maximizing "greenness threshold," discussed in the text. Figure 3.8(B) shows that a greenness threshold specifies both P(Hit) and P(False Alarm), and that the reverse is also true: a given value of P(False Alarm) specifies a corresponding greenness threshold. Thus this box solves for the rate-maximizing P(False Alarm). To simplify the notation, let $\phi = P$(False Alarm). Then P(Hit) $= \phi^k$, where k ($0 \leq k \leq 1$) determines the curvature of the ROC curve.

We are interested in how changes in P(False Alarm) affect the relationship between $\bar{e}$, the mean energy gain per encounter with green caterpillars, and $\bar{h}$, the mean handling time per encounter with green caterpillars:

$$\bar{e} = g_e \phi^k + b_e \phi$$
$$\bar{h} = g_h \phi^k + b_h \phi,$$

where

$$g_e = \frac{\lambda_g}{\lambda_g + \lambda_b} e_g \quad \text{and} \quad g_h = \frac{\lambda_g}{\lambda_g + \lambda_b} h_g$$

$$b_e = \frac{\lambda_b}{\lambda_g + \lambda_b} e_b \quad \text{and} \quad b_h = \frac{\lambda_b}{\lambda_g + \lambda_b} h_b.$$

To study the relationship between $\bar{e}$ and $\bar{h}$, we use differentiation

$$\frac{\partial \bar{e}}{\partial \phi} = g_e k \phi^{k-1} + b_e$$

$$\frac{\partial^2 \bar{e}}{\partial \phi^2} = -g_e k (1-k) \phi^{k-2},$$

and we study the relationship between ϕ and $\bar{h}$ with implicit differentiation

$$1 = g_h k \phi^{k-1} \frac{\partial \phi}{\partial \bar{h}} + b_h \frac{\partial \phi}{\partial \bar{h}};$$

thus

$$\frac{\partial \phi}{\partial \bar{h}} = \frac{1}{g_h k \phi^{k-1} + b_h}.$$

Differentiating again, we find that

$$0 = -g_h k (1-k) \phi^{k-2} \frac{\partial \phi}{\partial \bar{h}} + \frac{\partial^2 \phi}{\partial \bar{h}^2} (g_h k \phi^{k-1} + b_h);$$

thus

$$\frac{\partial^2 \phi}{\partial \bar{h}^2} = g_h k (1-k) \phi^{k-1} \left(\frac{\partial \phi}{\partial \bar{h}} \right)^2.$$

BOX 3.6 (CONT.)

By the chain rule

$$\frac{\partial \bar{e}}{\partial \bar{h}} = \frac{g_e k \phi^{k-1} + b_e}{g_h k \phi^{k-1} + b_h} > 0, \ \bar{e} \text{ always increases with } \bar{h}.$$

$$\frac{\partial^2 \bar{e}}{\partial \bar{h}^2} = -g_e g_h \left[k(1-k)\phi^{k-2} \frac{\partial \phi}{\partial \bar{h}} \right]^2 < 0, \ \bar{e} \text{ shows negative acceleration with } \bar{h},$$

and the rate we want to maximize (R) is

$$R = \frac{\bar{e}}{\tau + \bar{h}},$$

where τ is the expected time between encounters. Thus the marginal-value theorem can be used to find the rate-maximizing P(False Alarm), or ϕ^*. Figure B3.6 shows this solution: when τ is long (τ_1), a higher P(False Alarm) (ϕ_1^*) is acceptable; when τ is short (τ_2), a lower P(False Alarm)(ϕ_2^*) should be chosen.

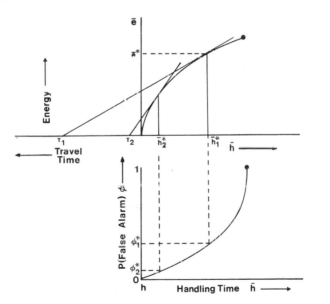

Figure B3.6 Solution for the rate-maximizing P(False Alarm). The relationship between $\bar{e}$ and $\bar{h}$ is shown in the upper panel, and the optimal pair ($\bar{h}^*$, $\bar{e}^*$) can be found by the usual tangent argument. The lower panel shows how $\bar{h}^*$ is related to the rate-maximizing P(False Alarm). The rate-maximizing P(False Alarm) is higher when travel time is longer.

mean net energy gained per encounter with the "system" and the mean involvement time with the system:

$$\bar{e} = \frac{\lambda_g}{\lambda_g + \lambda_b} \, e_g [P(\text{False Alarm})]^k + \frac{\lambda_b}{\lambda_g + \lambda_b} \, e_b P(\text{False Alarm})$$

$$\bar{h} = \frac{\lambda_g}{\lambda_g + \lambda_b} \, h_g [P(\text{False Alarm})]^k + \frac{\lambda_b}{\lambda_g + \lambda_b} \, h_b P(\text{False Alarm}),$$

where e_g and e_b are the respective energy values for good and bad profitability types, h_g and h_b are the handling times, and λ_g and λ_b are the encounter rates. Implicit differentiation shows that the relationship between $\bar{e}$ and $\bar{h}$ is negatively accelerated; thus the marginal-value theorem can be used to find the rate-maximizing $(\bar{h}, \bar{e})$ pair, and this result in turn specifies the rate-maximizing $P(\text{False Alarm})$ (see Box 3.6). As the time between encounters increases, the forager should choose larger values of $P(\text{False Alarm})$, specifying larger mean energy gains at the expense of longer mean handling times—at low encounter rates a rate-maximizer is less bothered about avoiding false alarms!

Up to this point, our discussion of recognition constraints has assumed only that certain inabilities to distinguish one thing from another constrain the forager. In Chapter 4 we discuss incomplete information problems that differ from the recognition constraint problems discussed here, because they suppose that the forager decides what price to pay for information. Chapter 4 asks more direct questions about information: how much is information worth, and should the forager go out of its way to get more information.

3.8 Conclusion

This chapter considers some simple changes in the basic formulations of foraging theory. These changes show the limitations and strengths of the basic models. Would-be testers might check that the constraints of their "forager" match those in the theory; if they do not, then the "changed constraint" models discussed here may tell the testers what differences to expect. In general, modifications of the constraints of the basic models have followed our "one change at a time" approach. An important problem for future theoreticians to study will be the possible interactions between these changes. The next chapter considers the most interesting and difficult constraint assumption: complete information.

3.9 Summary

This chapter considers six changes in the constraints of the basic prey and patch models: (1) simultaneous encounters, (2) the ability to encounter new items while handling old ones, (3) loosely clumped prey items, (4) travel restrictions and central-place foraging, (5) nutrient and toxin constraints, and (6) recognition constraints. Table 3.1 summarizes the results of each change.

Table 3.1
Summary of changed constraint models

Changed constraints	Which model is changed: prey or patch?	Results	Authorities
3.2 Simultaneous encounters allowed	Patch: prey choice within patches. The model treats a clump of simultaneously encountered prey like a patch with a discontinuous gain function. The modeler lists all possible tactics for using the clump such that the forager can choose only one tactic per clump.	(1) A less rewarding (smaller e) and less profitable tactic should never be preferred. (2) At short travel times (between clumps), more profitable tactics should be preferred. (3) At long travel times, less profitable but more rewarding (higher e) tactics should be preferred	Engen and Stenseth (1984); Stephens et al. (1986)
3.3 Search and handling are not mutually exclusive (overlapping encounters)	Patch	The forager should stay longer than the patch model predicts. Time spent in patches increases with the rate of overlapping encounters.	McNair (1983); Lucas and Grafen (1985)
3.4 Sequential encounter dependencies	Prey	(1) Prey ranking by profitability fails. (2) All encounter rates affect prey choice.	McNair (1979); Lucas (1983)
3.5 Travel restrictions Encounter at a distance	Prey	(1) Smaller prey are acceptable when close. (2) Any prey acceptable from far away must be acceptable when close.	Schoener (1979)

(Continued)

Table 3.1 (Continued)

Changed constraints	Which model is changed: prey or patch?	Results	Authorities
Single-prey loader	Patch: prey choice within patches	Similar to simultaneous encounter model. Size-selectivity increases with distance from the central place.	Orians and Pearson (1979); Lessells and Stephens (1983)
Multiple-prey loader	Patch	The basic patch model is extended to solve for the rate-maximizing load-size. Load size increases with distance from the central place.	Orians and Pearson (1979)
3.6 Nutrient and toxin constraints	Prey	Nutrient constraints predict partial preferences for unprofitable prey items. Toxin constraints can produce partial preferences for all prey.	Pulliam (1975); Belovsky (1978)
3.7 Recognition constraints			
Non-zero recognition time	Prey	Encounter rates with unprofitable types can affect their inclusion.	Charnov and Orians (1973); Elner and Hughes (1978); Erichsen et al. (1980); Houston et al. (1980)
Imperfect resemblance	Prey	The forager controls the extent of discrimination between two types by setting the probability of a false alarm. The rate-maximizing P(False Alarm) is high when habitat rates of intake are low.	Getty and Krebs (1985); Getty (1985)

4 Incomplete Information

4.1 Introduction

Perhaps the most common criticism of the early foraging models was that they assumed "complete information." The models of Chapters 2 and 3 assume that foragers behave as if they know encounter rates, profitabilities, and gain functions. In this chapter we ask, what is the rate-maximizing behavior when an animal must both acquire information and forage. Many authors view incomplete information as an unsolved problem in foraging theory (e.g. Pyke et al. 1977, Werner and Mittelbach 1981), but it is not unsolved because of a lack of effort. A surprising number of authors have studied incomplete information problems (Estabrook and Jespersen 1974, Bobisud and Potratz 1976, Oaten 1977, Arnold 1978, Krebs et al. 1978, Green 1980, 1984, McNamara and Houston 1980, Ollason 1980, Pulliam and Dunford 1980, Harley 1981, Iwasa et al. 1981, Killeen 1981, Orians 1981, Pulliam 1981, Houston et al. 1982, McNamara 1982, Lima 1983, 1985a, Stewart-Oaten 1983, Clark and Mangel 1984, Lester 1984, Regelmann 1984, Kacelnik and Krebs 1985). Why then, has so much work left the problem unsolved? The simple answer is that information is more than one problem (the more complex answer is left until the end of the chapter).

This chapter only discusses information in the basic prey and patch models of Chapter 2. These models represent only a few of the foraging problems animals face, but we focus our attention on these few because they have played a major role in the development of information models.

Information and ambiguity. Information problems usually follow the same pattern. Recall that the basic models assume a repetitive cycle of search-encounter-decide. In Chapters 2 and 3 (except in section 3.7), we assumed that foragers encountered types (of prey items or patches) that they recognized instantly. Following this use of the word "type," we propose two definitions. A forager recognizes *types* upon encounter, but a type may only represent a class of things (prey items or patches). A forager cannot recognize *sub-types* upon encounter. For an insectivorous bird, cones of

conifer trees may be a recognizable type, and the indistinguishable sub-
types might be fir cones versus pine cones, or cones with no insects in
them versus cones with ten insects in them.

The basic models assume that types are unambiguous, that is, types are
composed of only one sub-type. A simple information problem exists when
types are *ambiguous*, meaning that they are composed of more than one
sub-type. How should a long-term rate maximizer treat recognizable types
when it knows that they are divided into indistinguishable sub-types? We
divide this broad question into three categories: the value of prior recognition (section 4.2), tracking a changing environment (section 4.3), and
patch sampling (section 4.4).

The mathematical apparatus required to solve information problems is
called statistical decision theory. The principal concepts underlying statistical decision theory are (1) Bayes' Theorem and (2) conjugate distributions.
Box 4.1 reviews these ideas.

4.2 The Value of Recognition

In section 3.7 we saw how a rate-maximizer might work within recognition
constraints. Here we ask a slightly different question: How much is recognition worth? Gould (1974) has studied this question in the context of human
decision makers. We outline his analysis here because it helps to make
several points about information in general, and about prior recognition
in particular.

Imagine that some type is divided into k indistinguishable sub-types represented by $s_1, s_2, \ldots, s_k$. The forager may make some decision about
exploiting the type upon encounter, designated by the decision variable Y.
The forager's pay-off $[H(s_i, Y)]$ depends on its decision Y, and on the sub-
type that occurs s_i. A prior probability distribution $[p_1, p_2, \ldots, p_k$, where
$p_i = P(s_i), \Sigma\, p_i = 1]$ summarizes the forager's knowledge. If the forager has
no further knowledge, then it should choose the Y that maximizes the
average pay-off (Y^*), where

$$\sum_{i=1}^{k} p_i H(s_i, Y^*) = \max_{Y} \sum_{i=1}^{k} p_i H(s_i, Y). \tag{4.1}$$

However, if the forager can recognize sub-types, then it should use this *recognition* to choose a different Y value for each sub-type (Y_i^* for s_i). In
symbols,

$$H(s_i, Y_i^*) = \max_{Y} H(s_i, Y). \tag{4.2}$$

BOX 4.1 BAYES' THEOREM AND STATISTICAL DECISION THEORY

Statistical decision theory is an area of active research in mathematics, economics, and statistics, and it contains many unsolved technical problems. McNamara and Houston (1980) have written an excellent introduction to statistical decision theory as applied to animal behavior; outside animal behavior Raiffa (1968) provides a readable introduction. DeGroot's (1970) book is definitive, but difficult reading.

BAYESIAN STATISTICS

The forager's knowledge of the relative likelihood of the occurrence of each possible sub-type can be summarized by a *prior probability distribution*. A forager may know that half of all patches are full and half are empty. Upon encounter with a particular patch, the forager's "prior" distribution would be $P(\text{prey}) = P(\text{no prey}) = \frac{1}{2}$. Prior probabilities are sometimes called subjective probabilities, because they are based on preconceptions about the whole class of patches rather than on information about a particular patch.

Suppose that after entering a patch and searching there for a while the forager observes a rustle in the grass. This observation changes the forager's assessment of the likelihood that the patch contains prey. Bayes' theorem provides a method of updating prior information in the light of such an experience.

Bayes' Theorem:

$$P(\text{prey}\,|\,\text{rustling}) = \frac{P(\text{rustling}\,|\,\text{prey})\ \mathbf{P(Prey)}}{P(\text{rustling})} \qquad \text{(B4.1.1a)}$$

$$P(\text{no prey}\,|\,\text{rustling}) = \frac{P(\text{rustling}\,|\,\text{no prey})\ \mathbf{P(no\ prey)}}{P(\text{rustling})}. \qquad \text{(B4.1.1b)}$$

$P(\text{rustling})$ in the denominators equals the sum $P(\text{rustling}\,|\,\text{prey})\mathbf{P(prey)}$ $+ P(\text{rustling}\,|\,\text{no prey})\mathbf{P(no\ prey)}$, or the total probability of grass rustling. The boldfaced terms are the prior or subjective probabilities of full and empty, and the conditional probabilities on the left side of the equations are the *posterior* probabilities. This process can be repeated using the posterior probabilities as new prior probabilities, should new information come to light. Bayes' theorem provides a powerful description of information gain. The models presented in this chapter rely heavily on Bayes' theorem.

CONJUGATE DISTRIBUTIONS

We chose a simple example to illustrate Bayes' theorem. There may be many possible sub-types (even an infinite number). The calculation of the

BOX 4.1 (CONT.)

posterior probability distribution can be messy. Happily, mathematicians have already worked out the implications of particular types of experience in many important cases. As an example, suppose that a decision-making great tit begins to feed on winter moth larvae. The great tit's "decision" about whether to continue feeding on patches of these larvae depends on the mean larva size. Further, imagine that the great tit samples from a normal distribution of larval size with an unknown mean and a known variance, σ^2, and that the great tit's prior knowledge of the mean is summarized by another normal distribution with mean μ and variance s^2. After the great tit makes n measurements (it may measure by comparing the n larvae eaten so far to its bill length) and finding that their average is $\bar{X}$, the posterior distribution of the mean is a normal distribution with mean

$$\mu' = \frac{\dfrac{\mu}{s^2} + \dfrac{n\bar{X}}{\sigma^2}}{\dfrac{1}{s^2} + \dfrac{n}{\sigma^2}} \tag{B4.1.2a}$$

and variance

$$s^{2\prime} = \frac{1}{\dfrac{1}{s^2} + \dfrac{n}{\sigma^2}}. \tag{B4.1.2b}$$

The great tit's posterior estimate of the unknown mean is expression (B4.1.2a). There are three probability distributions in this scenario. The unknown feature of the world is a parameter of a stochastic process, that is, the mean of the normal distribution of larva size, the first distribution involved. The two remaining distributions are the prior and posterior distributions of the unknown mean. Results like these are called conjugate families of distributions, because the prior and the posterior distributions are from the same family—here normal prior, normal posterior.

These two expressions, (4.1) and (4.2), can be used to calculate how much the forager should be willing to pay for perfect prior recognition. The *value of recognition* is the difference

$$\sum_{i=1}^{k} p_i H(s_i, Y_i^*) - \sum_{i=1}^{k} p_i H(s_i, Y^*). \tag{4.3}$$

The left-hand term is the average pay-off given that the forager makes a different decision (Y_i^*) for each sub-type it encounters, and the right-hand term is the average pay-off when the forager must make the same decision regardless of which sub-type it encounters ($Y*$). Gould (1974) uses this definition to make several important and counter-intuitive points.

How does the prior distribution affect the value of prior recognition? Obviously, if all the p_i's but one are zero, then the value of prior recognition is nil. Why should the forager pay anything to find out what it already knows? Gould's second result is more surprising. The value of prior recognition is maximized when there are only two possible sub-types!

This outcome is not as puzzling as it may seem. Imagine that a patch may be any of three possible sub-types: 0 prey per patch, 5 prey per patch, and 100 prey per patch. Suppose that each of seven habitats has a different permutation of patch types (three with only one sub-type, three with two sub-types, and one with all three sub-types). The forager knows which habitat has which permutation of sub-types. Gould's result says that recognition is worth most *in one* of the three habitats where only a pair of sub-types exists, probably the habitat with the two sub-types 0 prey per patch and 100 prey per patch. Similar results come up in our discussion of patch sampling.

Gould defines the value of prior recognition in terms of the decision problem at hand, the pay-off function H, the prior distribution of sub-types, and so on. Other measures of the information value (e.g. DeGroot 1970, Shannon and Weaver 1949) treat all decreases in environmental ambiguity equally. The Shannon-Weaver index, for example, commits what might be called the academic fallacy: it assumes that certainty is intrinsically valuable. However, in foraging theory (and similar statistical decision problems) certainty is only valuable if it increases the forager's rate of energy intake. The value of information is generally finite, and partial reductions in ambiguity often may be good enough.

THE MARKET VALUE OF PREY RECOGNITION

To illustrate Gould's ideas, we consider a simple model of diet choice. Suppose that a forager encounters a single prey type, and that this type consists of two sub-types which have net energies e_1 and e_2, handling times h_1 and h_2, and encounter rates λ_1 and λ_2. These assumptions follow the conventional two-prey problem (see Chapter 2), except that we suppose that the forager cannot tell the sub-types apart *unless* it pays a discrimination cost of r_e calories and r_t seconds. We assume that if discrimination were free ($r_e = r_t = 0$), then the forager would ignore sub-type 2. To simplify the discussion we use the notation

$$R_{\{1\}} = \frac{\lambda_1 e_1}{1 + \lambda_1 h_1},$$

the rate achieved if discrimination is free, and

$$R_{\{1, 2\}} = \frac{\lambda_1 e_1 + \lambda_2 e_2}{1 + \lambda_1 h_1 + \lambda_2 h_2},$$

the rate achieved if prey are eaten indiscriminately. How much should the forager pay for recognition? This question can be answered algebraically by finding the set of (r_e, r_t) pairs that satisfies

$$\frac{\lambda_1 e_1 - (\lambda_1 + \lambda_2) r_e}{1 + \lambda_1 h_1 + (\lambda_1 + \lambda_2) r_t} > \frac{\lambda_1 e_1 + \lambda_2 e_2}{1 + \lambda_1 h_1 + \lambda_2 h_2}. \qquad (4.4a)$$

Rearranging, we find the equivalent expression

$$\frac{\lambda_2}{(\lambda_1 + \lambda_2)} [\lambda_1 e_1 h_2 - e_2(1 + \lambda_1 h_1)] > r_e(1 + \lambda_1 h_1 + \lambda_2 h_2) + r_t(\lambda_1 e_1 + \lambda_2 e_2).$$

$$(4.4b)$$

To show the relationship between r_e and r_t more clearly, we can rewrite this as

$$k > \alpha r_e + \beta r_t, \qquad (4.4c)$$

with k setting an upper limit on the price of discrimination. The forager will pay a lot for discrimination when e_2/h_2 is much smaller than $R_{\{1\}}$, and the forager will not pay much for discrimination when e_2/h_2 approaches $R_{\{1\}}$.

This model also shows how the energy costs of discrimination are traded off against the time costs of discrimination. At indifference (when expression 4.4 is an equality) a one-calorie increase in energy costs must be matched by a decrease in time costs of α/β or $1/R_{\{1, 2\}}$ seconds, and a one-second increase in time costs must be matched by a decrease in energy costs of $R_{\{1, 2\}}$ calories. This result foreshadows a technical point that comes up in Chapter 5. The rate of gain when there is no discrimination, or $R_{\{1, 2\}}$, is also the "marginal rate of substitution" of time for energy costs of discrimination (see Chapter 5 for discussion of marginal rates of substitution).

The fundamental information problem is, how do foragers value information? We presented the example above because there are good prospects for using discrimination cost models to study this question empirically. Experiments can be designed that give foragers the following options: find out at price x, or remain ignorant for free. Some empirical work (Elner and Hughes 1978, Erichsen et al. 1980, Houston et al. 1980, Getty and Krebs 1985) suggests simple ways in which recognition costs might be controlled experimentally. The question of the value of information may yield most easily to discrimination cost experiments and models (see also section 3.7).

4.3 Tracking a Changing Environment

This section analyzes a common foraging problem: how should a forager track a changing environment. Prey or patch types may change in quality with time. A prey or patch type that is not worth exploiting now may suddenly and unpredictably improve in quality. If factors such as season, time of day, or temperature do not give the forager "cues" to these changes, then the forager must sample; that is, it must periodically check its options to see if they have changed. A forager encountering a system of Batesian mimics may have to sample to determine whether it is traveling through a patch of models (bad prey) or mimics (good prey).

In this section we first analyze the "optimal tracking" policy, assuming that the forager can easily discriminate between good and bad items once it has sampled them, and then we present a model in which the difference between good and bad is more difficult to detect. In effect we separate the problem of tracking a change from the problem of detecting a change.

A RUN OF BAD LUCK WITH INSTANT RECOGNITION

This model focuses on how frequently a forager should check an unacceptable prey or patch type (for simplicity, we refer only to prey types from now on). There are four principal features of the model.

1. *A Varying Prey Type.* Called type X, this varying prey type consists of two sub-types, "bad" and "good." They might, for example, be a model and its Batesian mimic (Estabrook and Jespersen 1974). The notations v_g and v_b denote the value of good and bad sub-types, respectively. The forager can easily distinguish one sub-type from the other after sampling them: it can tell immediately after consuming an item whether the item was good or bad.

2. *The Alternative.* When the forager encounters an X, the forager may ignore it and take another option, called the "alternative," which has value v_a. The alternative's value is intermediate, $v_g > v_a > v_b$.

3. *Time per Encounter.* Time is measured in encounters with X. Maximizing the expected gain per encounter maximizes the long-term rate of energy intake, because the expected time per encounter is the same regardless of whether the forager attacks good, bad, or alternative prey types.

4. *Relationship between Present and Future Sub-Types.* The present sub-type is related to future sub-types in the following way. The probability that the sub-type will stay the same from one encounter to the next is a

constant. Symbolically,

$$P(\text{Good at } i + 1 \,|\, \text{Good at } i) = P(\text{Bad at } j + 1 \,|\, \text{Bad at } j) = q,$$

where q is the *probability of a repeat*, and $1 - q$ is the *probability of a change*. The notation q characterizes the stability of the varying system. The average length of a run of a given type (i.e. good or bad) is an increasing function of q, $E(\text{run length}) = q/(1 - q)$.

THE FORM OF THE SAMPLING RULE

A first order Markov chain controls the transitions between good and bad. In a first order Markov chain the prior (or subjective) distribution of states at time j depends solely on the state at $j - 1$ (in a second order process it would depend on both states $j - 1$ and $j - 2$). This means that observing two "bads" in a row provides the forager with no more information than observing one. Every bad observed tells the forager the same thing. Therefore the tracking rule is simple: Take X's as long as "goods" are observed; after any bad is observed ignore X's (and take the mediocre alternatives) for N encounters, and sample at the $N + 1$st encounter (Bobisud and Potratz 1976 and Stephens 1982 present more detailed justifications of this rule). Here N is called the sampling period, because its inverse is the sampling frequency. The sampling period (N) is the model's decision variable.

A GRAPHICAL ARGUMENT

Figure 4.1 shows a run of bad luck and the behavior of two hypothetical foragers, one adopting a sampling period (N) of four and the other adopting a sampling period of twelve. Consider the behavior of an omniscient forager that begins to eat alternatives just before observing the first bad, and switches back to goods immediately when the run of bad luck ends. In comparison to this omniscient behavior the constrained foragers make two kinds of errors. First, they make sampling errors (marked S in Fig. 4.1); they take a bad when they could have taken a better "alternative" prey. Second, they make overrun errors (marked O in Fig. 4.1); they accept the mediocre alternatives even though the varying system has changed back to good. The frequent sampler ($N = 4$) makes many sampling errors but few overrun errors; the infrequent sampler ($N = 12$) makes few sampling errors but many overrun errors.

An optimal sampler must balance the costs of sampling too frequently, and taking too many bads, against the potential loss of opportunity in missing the change back to good. Where the rate-maximizer strikes this balance depends on the relative costs of sampling and overrun errors. In this model a single sampling error costs $v_a - v_b$, and a single overrun

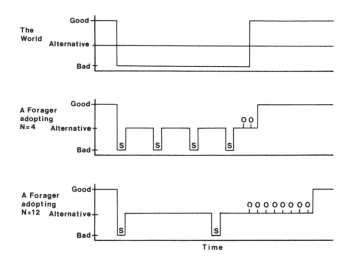

Figure 4.1 A sequence of good and bad sub-types of the varying type X is shown. The effects of two sampling periods ($N - 4$ and $N = 12$) are shown. There are two types of errors, sampling errors marked S and overrun errors marked O. When the sampling period is short, there are many sampling errors but few overrun errors. When the sampling period is long, there are many overrun errors but few sampling errors.

error costs $v_g - v_a$. The ratio of the costs of these two errors,

$$\varepsilon = \frac{v_a - v_b}{v_g - v_a} = \frac{\text{cost of a sampling error}}{\text{cost of an overrun error}}$$

(ε is called the error ratio), and the varying prey type's probability of a repeat (q) together determine the rate-maximizing sampling period.

Before discussing the optimal tracking policy, let us consider a forager that always takes the varying prey type or always take the alternative prey type, whichever is the best "on average." Obviously the average value of the alternative is v_a. What is the average value of the varying prey type X? In the long run a forager that always attacked the varying prey type would receive equal numbers of bads and goods, because the process that describes the change is symmetric [i.e. $P(\text{Good at } i + 1 | \text{Good at } i) = P(\text{Bad at } j + 1 | \text{Bad at } j)$]. In the jargon of stochastic processes, one says that the *equilibrium* probability of a good (equals the equilibrium probability of a bad) equals one-half. The average yield from the varying prey type is $(v_g + v_b)/2$. The alternative is better on average than the varying type if $v_a > (v_g + v_b)/2$. A little algebra shows that this is equivalent to $\varepsilon > 1$ (Fig. 4.2[A]). The *average* yield from the varying prey type is independent of q, so the non-tracking forager should always choose the varying option when $\varepsilon < 1$ and the alternative when $\varepsilon > 1$ (Fig. 4.2[A]).

Now we turn to the conditions in which sampling pays. By using results from the theory of stochastic processes, we can find more complicated equilibrium probabilities. The notation $p_g(N, q)$ is the equilibrium probability that a forager, adopting a sampling period of N when the probability of a repeat equals q, will eat a good; $p_b(N, q)$ is the corresponding probability of eating a bad, and $1 - p_b(N, q) - p_g(N, q)$ is the equilibrium probability of eating an alternative. [Estabrook and Jespersen 1974 ele-

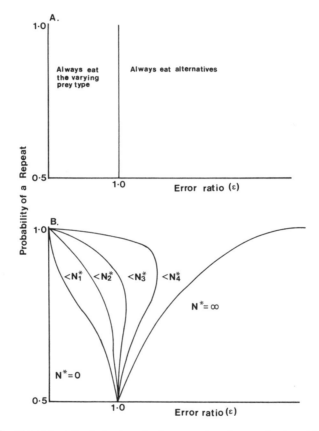

Figure 4.2 (A) shows the behavior of a forager who always attacks whichever prey type is best "on average." Such a forager's behavior is independent of the probability of a repeat, q. (B) shows the sets of q and ε where particular sampling periods are optimal. Notice the two non-tracking regions N^* equals zero (always take the varying prey type) and N^* equals infinity (always take the alternative). The regions marked N_i^* are regions where there is a range of optimal intermediate sampling periods, $N_4^* > N_3^* > N_2^* > N_1^*$. Each N_i^* represents a range, not a single value.

gantly derived expressions for $p_b(N, q)$ and $p_g(N, q)$]. The expected gain per encounter is

$$v_b p_b(N, q) + v_g p_g(N, q) + v_a[1 - p_b(N, q) - p_g(N, q)], \qquad (4.5a)$$

or

$$(v_g - v_a)[p_g(N, q) - \varepsilon p_b(N, q)] + v_a. \qquad (4.5b)$$

Because $(v_g - v_a)$ is a positive constant maximizing

$$p_g(N, q) - \varepsilon p_b(N, q) \qquad (4.5c)$$

also maximizes the expected gain per encounter. This expression is only a function of N, q, and ε, as we argued that it would be.

The techniques for finding the rate-maximizing sampling period (N^*) are not important here (see Stephens 1982, 1987). A graphical solution is found by turning the problem on its head and asking, for example, given that $N^* = 10$ what must ε and q be? Figure 4.2(B) shows the sets of ε and q that give various values of N^*. Over a wide range of ε and q values, N^* equals either zero (no tracking, attack the varying prey type regardless of experience) or infinity (no tracking, attack the alternative regardless of experience). Where $q = \frac{1}{2}$, observations have no information value, and the best tactic is to stick with the option that has the highest average pay-off as described above. The conditions, between N^* equals zero and N^* equals infinity, where tracking is an economically sound policy are narrow. (However, if the figure plotted ε against expected run length —a nonlinear relative of q—this narrowness would not be so striking.) Figure 4.3(A) shows a graph of the optimal sampling period N^* versus the error ratio (a slice of Fig. 4.2[B] at constant q). One might intuitively expect this increasing trend, because as sampling errors become relatively more expensive the sampling frequency ($1/N^*$) decreases.

Figure 4.3(B) shows a more surprising result: the probability of a repeat (q) plotted against the optimal sampling period N^* (a slice of Fig. 4.2[B] at constant ε). One might expect that as the average run length gets longer (large q) the sampling period should increase to match the average run length. However, Figure 4.3(B) shows that when $\varepsilon > 1$ long sampling periods occur at both high and low probabilities of repetition, with the shortest sampling period occurring at intermediate values of q. At high probabilities of repetition (long average run lengths) the optimal sampling period is long simply to match the long runs in the habitat, but at low probabilities of repetition (short average run lengths) conscientious tracking would require short sampling periods. However, frequent sampling does not pay at low probabilities of repetition, because knowing that a short-lived run is "on" is not a valuable piece of information. This

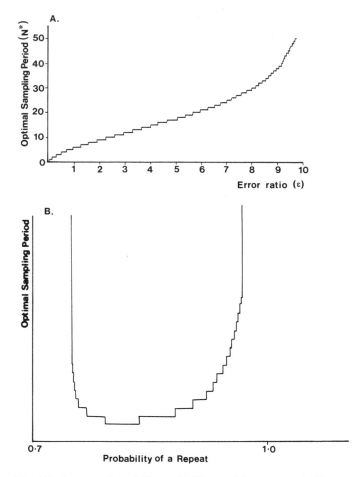

Figure 4.3 (A) shows a slice of Figure 4.2(B) parallel to the ε-axis. The graph gives the optimal sampling period as a function of the error ratio ε, at a constant probability of repetition ($q = 0.95$). (B) shows a slice of Figure 4.2(B) parallel to the q-axis. The graph gives the optimal sampling period as a function of the probability of a repeat at a constant error ratio ($\varepsilon = 2$).

is counter-intuitive. Tracking is a problem in the first place because habitats change, yet tracking is less valuable when habitats change frequently.

SUMMARY FOR TRACKING WHEN DISCRIMINATION IS EASY

There are three important conclusions.

1. *The Error Ratio Result.* There is a trade-off between the costs of sampling and the costs of opportunity loss from an overrun. A forager

that samples more will avoid overruns at the cost of excessive sampling; a forager that samples less will avoid costly sampling but will risk costly overruns.

2. *Tracking Is Not Always a Good Policy.* There is a wide range of conditions of low habitat "stability" (low q) and extreme values of the error ratio in which no tracking should occur even though the forager's experience still provides some information about the state of the environment.

3. *Habitat Stability Affects the Value of Information.* The need to sample frequently to "keep track" of a frequently changing habitat is counterbalanced by the reduced value of information about changes that are bound to be short-lived.

A more complicated model might capture more of the real world's details, but it would probably not change any of the qualitative results above. Moreover, because of its simplicity this model suggests experimental studies. Arranging experiments in which the somewhat restrictive assumptions of this model apply should be straightforward. Shettleworth et al. (in preparation) have carried out a laboratory experiment in which the varying and alternative prey types are represented by key in a Skinner box. They have altered the error ratio by manipulating the value of the alternative (v_a) and the value of the good sub-type (v_g). Their results qualitatively agree with the model.

DETECTING A CHANGE: A DISCRIMINATION PROBLEM

Two of the previous model's assumptions (a first order Markov transition rule and the forager's ability to immediately distinguish the varying prey type's sub-type) increase the information value of sampling. When either assumption is relaxed, a single sampling event cannot establish whether the varying prey type is good or bad. Instead, each sampling event will only affect the forager's "estimate" of the likelihoods of good or bad sub-types. This section focuses on the problem of recognizing a change.

To show the effects of violating the assumption of "immediate recognition," we consider one of McNamara and Houston's (1980) models. Suppose that the forager attacks a varying prey type like the one discussed above, except that the *good sub-type* is a mixture of desirable and undesirable items, and that the proportion of desirable items is p. Moreover, suppose that the *bad sub-type* is composed of 100% undesirable items. The model assumes that desirable and undesirable items can be immediately distinguished after being eaten, but because the good sub-type contains some undesirable items the sub-type of the varying system cannot be perfectly distinguished. Finally, assume that the varying prey type starts out in the good sub-type and switches *permanently* to the bad sub-type, according to the same Markov transition rule used above.

If the forager observes a desirable item, then it knows that the sub-type must be good because no desirable items occur in the bad sub-type. Thus the forager needs to know only the number of undesirable items observed since the last desirable item (McNamara and Houston 1980). What does a run of 10 undesirable items in a row mean? How many "undesirables" must be observed before the forager is convinced that the varying prey type has turned for the worse? Bayes' theorem (Box 4.1) can be used to reach some qualitative conclusions about the meaning of a run of bad luck, $P(\text{No Switch} \mid x \text{ undesirables}) =$

$$\frac{P(x \text{ undesirables} \mid \text{No Switch})P(\text{No Switch})}{P(x \text{ undesirables} \mid \text{No Switch})P(\text{No Switch}) + P(x \text{ undesirables} \mid \text{Switch})P(\text{Switch})}.$$

The prior probability that the sub-type has not changed after x undesirables is q^{x+1} (the sub-type must stay good on each of its $x + 1$ opportunities to change). If the sub-type is good, then the probability of observing x undesirables in a row is $(1 - p)^x$; therefore $P(x \text{ undesirables} \mid \text{No Switch})P(\text{No Switch}) = (1 - p)^x q^{x+1}$. The term $P(x \text{ undesirables} \mid \text{Switch})P(\text{Switch})$ is a little more complicated. To find this term, we break the event "Switch" into the $x + 1$ points before the xth undesirable was observed at which the change to bad might have occurred. The probability that the good to bad transition occurs at the ith opportunity is $(1 - q)q^i$ (the geometric distribution). The probability that the forager observes x undesirables if the switch occurs at j ($j < x$) is $(1 - p)^j$, since j undesirables must occur while the sub-type is still good. Therefore

$$P(x \text{ undesirables} \mid \text{Switch})P(\text{Switch}) = (1 - q)\sum_{0}^{x}(1 - p)^i q^i.$$

By substitution, we find that the probability that the sub-type is still good after x undesirable items have been observed is

$$\frac{(1 - p)^x q^{x+1}}{(1 - p)^x q^{x+1} + (1 - q)\sum_{0}^{x}(1 - p)^i q^i}. \tag{4.6}$$

Figure 4.4 shows the posterior probabilities that the varying type is still good after the forager observes a given number of undesirable items. When the probability of a repeat is low, the forager's "confidence" is more easily shaken by a run of undesirables. The probability of a desirable item when the sub-type is good (p) shows a more interesting result: if the p is low,

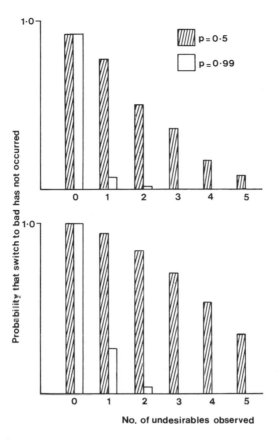

Figure 4.4 Both panels show the probability that the switch to bad has not occurred as a function of the number of undesirable items observed so far. The shaded bars are calculated on the assumption that p, the probability of a desirable item when the sub-type is still good, is 0.5. The open bars assume that $p = 0.99$. The upper panel assumes that the probability of a repeat is 0.900. The lower panel assumes that the probability of a repeat is 0.975. (Calculations following McNamara and Houston 1980.)

then it takes longer to convince the forager that a run of bad luck is more than a random fluctuation. When p is low, good and bad are more alike. This general "comparison effect" is intuitively appealing; similarities between the two sub-types make discrimination more time-consuming and difficult. McNamara and Houston (1980) used this model to explain a frequently described effect from the animal learning literature, the partial reinforcement extinction effect (PREE). A pigeon may be trained to peck a key for a reward on $100p\%$ of its pecks, and after a long period of training the rewards are abruptly stopped by the experimenter. If p is low, then it takes longer for the pigeon to stop responding (behavioral extinction); if p is high, then extinction is fast. This is true even though pigeons pecking for 10% rewards can be trained to peck at the same rate as pigeons pecking for 90% rewards. This effect has presented difficulties for some simpler theories of feeding behavior (e.g. linear operator models of learning, Kacelnik and Krebs 1985), but it fits qualitatively into an incomplete information model. In contrast to the models psychologists develop to account for the PREE (e.g. Capaldi 1966), this model is not derived as a description of the data but from *a priori* considerations. Although it predicts the trend qualitatively, preliminary observations suggest that animals may be more persistent than information grounds alone lead us to expect (Kacelnik and Krebs 1985).

This model is incomplete. Although it makes statements about what the forager knows after X undesirable items, it says nothing about how this knowledge can be put to use. A complete solution would find an optimal stopping rule, and such a model must take explicit account of the value of the forager's alternatives. A solution will not be attempted here, but the reader might compare this problem with the detection problem studied by Krebs et al. (1978).

Imagine the difficulties of adding this type of detection problem to the problem of finding an optimal sampling period. We suspect that making discrimination more difficult would tend to make conditions for tracking even narrower, because it would make each sampling event less informative. Another probable effect would be a change in the pattern of sampling. The forager might persist for a long time when the varying prey type first goes bad, but make short and cursory samples after it is "convinced" of a turn for the worse. The assumptions that make up any model represent a trade-off between those which make the model a simple guide and those which may make it realistic but clumsy. We have argued that tracking can be logically separated from detection. It is important to have models that combine these elements, but this should not hinder the reasonable experimental task of studying these elements independently.

4.4 Patch Sampling

ANOTHER REASON TO MOVE ON

The deterministic marginal-value theorem views patch depression (the "marginal" decrease of the within-patch rate of gain) as the principle reason for moving on to another patch. Patch depression may occur for many reasons (Charnov et al. 1976), but one of the most important of these is that as the forager spends more time in a patch, it also spends more time revisiting previously searched ground. Some foragers, such as woodpeckers moving along tree trunks, may experience little or no patch depression because they can search systematically, and according to the deterministic marginal-value theorem, these foragers should either always ignore a patch of a given type or always search it exhaustively (see also section 9.5). If there is no depression, then there is no reason to give up before emptying the patch.

However, a forager may leave a partially full patch for another reason: patch sampling or assessment. A forager may leave before completing an exhaustive search, because its foraging in the patch tells it that the patch is an "inferior" sub-type (Lima 1983, 1985a, Kacelnik and Cuthill in press).

A SIMPLE MODEL

In a pioneering study of patch sampling Lima (1983) drilled 24 shallow holes in each of 60 short lengths of tree trunk (patches). Each hole may or may not have contained at least one piece of sunflower seed, and an opaque piece of masking tape covered each hole so that no visual cues could be used to tell patches containing many seeds from patches containing no seeds. Lima hung these artificial patches in a wood lot and trained downy woodpeckers (*Picoides pubescens*) to feed from them. The woodpeckers learned to pierce the masking tape to examine the contents of a hole, and they searched nearly systematically.

In Lima's experiments there was only one type of patch (since patches were externally identical), but there were always two sub-types. The two patch sub-types were (1) EMPTY sub-type: in one-half of the patches there were no seeds, and (2) p-FULL sub-type: in one-half of the patches a proportion p of the holes contained seeds. Lima's experimental treatments were $p = 1$, $p = \frac{1}{2}$, and $p = \frac{1}{4}$. The forager's problem here is formally similar to the problem of "detecting a change" discussed above. Once the forager finds a single seed, it "knows" (or should know) that the patch is full. Lima, therefore, argues that a sensible leaving rule has the form: "Leave a patch if no seeds are found by the nth hole opened; search the patch exhaustively if at least one seed is found by the time the nth hole is opened." The modeler's problem is to find the value of n (n^*) that maximizes the long-term rate of energy gain.

The problem analyzed here is simpler than Lima's because we let the p that characterizes the p-FULL sub-type be the probability that a hole is full in a p-FULL patch. In Lima's experiments if p was one-half, then there were 12 full holes in every p-FULL patch; in our version 12 holes is only the expected number of full holes per p-FULL patch. As usual, we want to choose the n that maximizes the long-term rate of energy gain:

$$R = \frac{E(G|n)}{\tau + E(T|n)},$$ (4.7)

where τ is the travel time between patches. $E(G|n)$ is the expected number of prey gained per patch given that n is the giving-up rule, and $E(T|n)$ is the expected time spent per patch given that n is the giving-up rule. The two possible actions (give up after n or search exhaustively) and the two patch sub-types (EMPTY or p-FULL; see Table 4.1) determine four possible

Table 4.1

Expected gains and times from possible tactics

| | | Patch sub-type | |
		Empty	Full
Action	Leave after n	$E(G) = 0$ $E(T) = n$	$E(G) = 0$ $E(T) = n$ (Error!)
	Search exhaustively	$E(G) = 0$ $E(T) = 24$ (Error!)	$E(G) > 0$ $E(T) = 24$

outcomes. A forager may "conclude" (by leaving before searching the patch exhaustively) that a patch is EMPTY. If it reaches this conclusion when the sub-type is really p-FULL, then it commits an error. Alternatively, if it concludes that the patch is p-FULL (by searching it exhaustively) when the patch is really EMPTY, then it commits the complementary error.

Table 4.1 shows the expected gain $[E(G)]$ and the expected time $[E(T)]$ that result from each of the four possibilities. The expected energy gain is zero in every case except "search exhaustively when the patch is p-FULL." The expected gain here is higher than $24p$ because it is the gain *given* that at least one item is found on or before n. This value can be found from the relationship

$E(G|$something by $n)P($something by $n)$

$+ E(G|$nothing by $n)P($nothing by $n) = 24p,$

and by using a little algebra we find that

$$E(G|\text{something by } n) = \frac{24p - (24 - n)p(1 - p)^n}{1 - (1 - p)^n}.$$ (4.8)

The calculation above is performed assuming that the patch is of sub-type p-FULL. The time in a patch can only be n or 24. The expected time per patch is just the weighted mean

$$E(T|n) = \frac{n}{2}[1 + (1 - p)^n] + \frac{24}{2}[1 - (1 - p)^n]. \qquad (4.9)$$

By substitution, we find the long-term rate of gain as a function of the giving-up rule n:

$$R = \frac{24p - (24 - n)p(1 - p)^n}{2\tau + (n + 24) + (n - 24)(1 - p)^n}. \qquad (4.10)$$

Since n can only take the discrete values 1, 2, 3, ..., tabulating the values of $R(n)$ shows the rate-maximizing n. Table 4.2 shows the results of such

Table 4.2

Optimal number of holes to check
before giving up (subjective probability
that the patch is EMPTY at leaving)

	Travel time	
p in p-FULL patches	1 time unit	100 time units
0.25	10 holes (0.97)	14 holes (0.98)
0.50	5 holes (0.97)	7 holes (0.99)
0.75	3 holes (0.98)	4 holes (0.996)
1.00	1 hole (1.00)	1 hole (1.00)

a tabulation. Notice the following two results: (1) in general, more holes should be opened for lower p values (analogous to the "comparison effect" in the previous section) and (2) even without patch depression our hypothetical woodpecker should be more tenacious when the between-patch travel time increases (Kacelnik and Cuthill 1986; see also Chapter 9). In general, the forager should be *at least* as tenacious for longer travel times or increased travel costs as it is for shorter travel times (McNamara 1982).

A strident advocate of the marginal-value theorem might claim that even if there is no physical patch depression, there is depression of subjective patch value. When the assessing forager finds that the next patch is more difficult to reach, it should be willing to tolerate a smaller likelihood that the present patch is good before leaving. As Table 4.2 shows, when

p equals 0.25, our hypothetical forager's assessment of patch value (likelihoods of EMPTY and p-FULL after 10 empty holes have been examined is the same regardless of whether the between-patch travel time (τ) is 1 time unit or 100 time units; yet when τ is 100, the forager must see 4 more empty holes before leaving. This post hoc–ism explains the phenomenon intuitively, but it does not detract from the important point that physical patch depression and patch assessment should share the limelight in economic analyses of patch-leaving decisions, and that most empirical studies to date have not tried to separate these factors.

For systematic searchers, assessment is probably the principal reason for patch leaving, but it is also important for another type of forager. Many foragers not only experience little patch depression, but they also never exhaust their patches. Such creatures would be sedentary—or partially sedentary—trap-builders, filter-feeders, and sit-and-wait foragers: barnacles, ant lions, net-building caddisfly larvae, web-building spiders. If a web-building spider finds itself in a site with a healthy procession of insect prey, then the spider's own activities may not reduce the arrival rate of prey significantly. If this is true, then there is no "marginal-value" reason to move on. The site-leaving behavior of partially sedentary foragers probably involves two components: (1) initial searching for an acceptable site (statistical decision theorists have studied such "optimal stopping" or "search" problems extensively; see DeGroot 1970) and (2) leaving after external and unpredictable conditions have made a previously acceptable site unacceptable. Although some authors (e.g. Townsend and Hildrew 1980) have applied the marginal-value theorem to site-leaving by sedentary hunters, an assessment model is more appropriate; the marginal-value theorem can apply only if site quality is continuously *and predictably* reduced as more time is spent on the site.

THE "FAILURE" OF THE MARGINAL-VALUE THEOREM

Much of the literature on patch assessment has compared it to the marginal-value *theorem*. Without exception this literature concludes that the marginal-value theorem fails (does not maximize rate) when patch assessment is important (Oaten 1977, Green 1980 and 1984, McNamara 1982). This is true only in a limited sense. These papers compare optimal patch assessment policies to those of a hypothetical animal that decides when to leave a patch using the following marginal-value *rule:*

Step 1. Measure the instantaneous rate of gain when foraging in a patch.
Step 2. Compare this rate with the optimal rate achievable in the habitat.

Step 3. Leave when the measured instantaneous rate drops to the level of the habitat rate of gain.

It can be unequivocally shown that when the forager's assessment of patch value changes as a result of experience in the patch, then this marginal-value *rule* does not result in long-term rate maximization.

This observation requires a retreat by the more ardent proponents of the marginal-value theorem. For example, it has occasionally been suggested that animals "should" somehow estimate the instantaneous rate of gain to determine their leaving policy (e.g. Pyke et al. 1977). Such a marginal-value rule *may* work when within-patch foraging has no information value, but it does not work when patch assessment is important. However, even when patch assessment is unimportant, this marginal-value rule still presents many unnecessary difficulties: how does the forager measure the instantaneous rate when prey come in discrete lumps?

The marginal-value *theorem* survives not as a rule for foragers to implement, but as a technique that finds the rate-maximizing rule from a known set of rules. For example, in Chapter 3 we used the logic of the marginal-value theorem (constructing tangents, etc.) to find which of two mutually exclusive prey items should be chosen. The resulting optimal rule has the form: choose the big prey item when $\lambda < k$, and choose the small prey item when $\lambda > k$ (where k is a constant; see Box 3.1). It would be a vain exercise to try to use a marginal-value *rule* to predict this simple patch-use behavior. However, the inapplicability of the marginal-value *rule* does not detract from the marginal-value *theorem's* central role in the solution, since we used the marginal-value theorem to find k. The marginal-value theorem is not a patch-leaving rule (we discuss patch-leaving rules in Chapter 8). It is a method for finding the best (long-term rate-maximizing) rule from a known set of possible rules. In this sense the marginal-value theorem works regardless of whether there is patch assessment. It can be used to solve all the examples McNamara (1982) presents to disprove the marginal-value *rule*. Green (personal communication) has recently shown that this is also true of his patch assessment model.

What Is Wrong with the Marginal-Value Rule?

Is there a general patch-leaving rule? McNamara (1982) has proposed such a rule, together with an enlightening analysis of how and why the marginal-value rule fails. Without giving the mathematical details, we present the highlights of McNamara's analysis.

The potential function. Our discussion of McNamara's results requires some formal definitions. The notation used here differs from McNamara's, because his notation differs from the notation we use elsewhere. Let t be the time spent so far in a patch. Let $\mathbf{X}_t$ be a vector that summarizes the forager's experience in the patch up to time t; $\mathbf{X}_t$ is called the state at time t. It might contain information about whether the grass has rustled and, presumably, about how many prey have been captured by time t. The experiences that contribute to $\mathbf{X}_t$ will be at least partially random, so $\mathbf{X}_t$ is a random variable. We can find a future expectation for any function of state and for any time u greater than the present time t. Symbolically,

$$E[f(\mathbf{X}_u)|\mathbf{X}_t = \mathbf{x}].$$

(We follow the convention of using capital letters to represent random variables and the corresponding small letters to indicate specific realizations of the random variables.) Let $G(\mathbf{x})$ be the function of the forager's "state" that translates it into energy gain. Any patch-leaving rule that the forager adopts will specify a patch residence time T. A particular decision rule defines the residence time T as a random variable. R^* is the maximum long-term rate of energy intake in the habitat.

For a given patch-leaving rule, and its associated residence time T, the expected gain from now (t, with $\mathbf{X}_t = \mathbf{x}$) until leaving is

$$a_T(\mathbf{x}) = E[G(\mathbf{X}_T)|\mathbf{X}_t = \mathbf{x}] - G(\mathbf{x}), \tag{4.11}$$

and the expected time left on the patch is

$$b_T(\mathbf{x}) = E(T|\mathbf{X}_t = \mathbf{x}) - t. \tag{4.12}$$

The difference,

$$a_T(\mathbf{x}) - R^*b_T(\mathbf{x}), \tag{4.13}$$

is the expected future gain on a patch minus the expected loss due to lost time. McNamara calls the maximum of this difference the potential function, $h(\mathbf{x})$:

$$h(\mathbf{x}) = \max_T[a_T(\mathbf{x}) - R^*b_T(\mathbf{x})]. \tag{4.14}$$

The patch residence time T^* at which this difference is maximized is the rate-maximizing patch residence time given that the state is $\mathbf{X}_t = \mathbf{x}$. This function is a list of states ($\mathbf{x}$) and the corresponding potential of each state. The rule McNamara proposes is: Stay, if the potential is positive; leave, if the potential drops to zero. This may seem circular, since one would often have to find T^*, the optimal patch residence time, to specify the potential. However, McNamara's model finds a general patch-leaving rule, not the optimal residence time; thus his rule is not circular. It seems plausible that a forager may be able to associate states with potential values. The potential will not change smoothly with changes in experience. Figure 4.5 shows how a typical potential function might change with experience (state). In the figure the potential jumps upward every time the forager captures a prey item, because a capture changes the forager's assessment of patch value for the better. (Note the similarity between the potential function and various rules of thumb for patch leaving discussed in Chapter 8).

The potential function bears a strong resemblance to the marginal-value rule, and it is possible to pick the potential function apart to show the relationship between the two. McNamara compares the potential function with the marginal-value rule in the special case in which the forager only gains information when it captures prey ("information by rewards").

Figure 4.5 An illustration of how a typical potential function might behave. When a capture occurs, the potential jumps to a higher value. In the interval between captures the potential steadily declines.

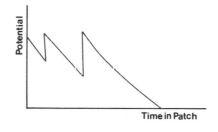

McNamara rewrites the potential function rule, for this special case, as follows. Leave the patch when

$$r(t^*) + p(t^*)\Delta(t^*) = R^*,$$

where $r(t)$ is a stochastic instantaneous rate of gain. It is an expected gain calculated using posterior likelihoods, measured over an infinitesimally small time interval, and divided by the same small time interval. It is defined independent of the forager's patch-leaving rule. In this expression, $p(t)$ is the probability of finding a prey item in the next instant, $\Delta(t)$ is the increment the potential would jump if a prey item were found in the next instant, and R is the optimal long-term rate of gain in the habitat. If the term $p(t^*)\Delta(t^*)$ were zero, then this leaving policy would be a stochastic version of the marginal-value rule. If the forager knew everything about patch quality, then $p(t^*)\Delta(t^*)$ would be zero because no experience could change the potential. However, when patch qualities are incompletely known, then $p(t^*)\Delta(t^*)$ will be positive, and the marginal-value rule will not work. The size of $p(t^*)\Delta(t^*)$ is a measure of how far off the marginal-value rule is. The instantaneous rate of gain at leaving, if the optimal leaving rule is followed, will be less than the habitat rate of gain (R^*) because, even when the instantaneous rate of gain has fallen to R^*, "by staying a further short time something could happen to make it optimal to stay" (McNamara 1982).

The spread result. McNamara's elegant comparison illustrates that when the spread (or variance) of the prior distribution of patch sub-types increases, a forager using the marginal-value rule does increasingly poorly compared with a forager following the optimal assessment policy. Consider a problem in which there are three patch sub-types, 0 prey per patch, 5 prey per patch, and 10 prey per patch. Compare the increase in potential resulting from capturing the first prey item in two cases. In the low-spread case we suppose that the three sub-types are equally likely. Here the first prey capture means that the patch may contain either 4 or 9 more prey. If the spread is increased by eliminating the mediocre patch sub-type (5 prey per patch) and making the 0 and 10 prey per patch sub-types equally likely, then the first prey capture provides a much more valuable piece of information. There are 9 prey left. By applying this logic to McNamara's formulation, we can see that when there is more spread, the jump in potential caused by a capture will be greater. The spread result is a close relative of Gould's (1974) two sub-type result (section 4.2), because when the distribution of sub-types is discrete, then the highest variance occurs when there are just two sub-types.

Oaten (1977) has shown that, by increasing the spread of sub-types, the marginal-value rule can be made to do arbitrarily poorly in comparison with the optimal assessment rule. This result points out an interesting problem in the patch assessment literature. Most assessment models have assumed that there are no external features that the forager might use to categorize patches into types before foraging in them. In nature, however, one might expect extreme spreads in patch quality to be reflected externally. A pinecone holding 10,000 pupae will not look like a pinecone holding 10. (A pinecone holding 10,000 pupae may seem ridiculous, but this is the kind of extreme that "arbitrary increases" in spread would require.) The interaction between external recognition and within-patch sampling may often be important in nature. Pyke (1981b) found that the flowers of monkshood contained the same distribution of nectar regardless of the number of flowers per inflorescence. Nevertheless, he found that bumblebees were more tenacious on large inflorescences than they were on small inflorescences. The value of finding a good big inflorescence (big is an externally recognized feature) is clearly greater than the value of finding a good small inflorescence. The difference in tenacity results from the interaction between external recognition and within-patch assessment. Neither Oaten's nor Green's (Oaten 1977, Green 1980, 1984) assessment models can account for this effect of inflorescence size.

Spread versus comparison. The spread result and the comparison effect are strange bedfellows. The spread result says that when sub-types are very different, it is important to tell them apart. The comparison effect says that when sub-types are very different, it is easy to tell them apart. The spread result and the comparison effect complement one another. When the cost of information is high, it is less valuable. When the cost of information is low, it is more valuable. This suggests that simple rules of thumb may work well as assessment strategies. When the spread is high, a simple rule (a single taste, a single peek) will often be enough to determine the sub-type. When the spread is low, a simple rule may lead to incorrect conclusions, but these misjudgments will not be too harmful. Elaborate and expensive sampling rules will be most useful at intermediate levels of sub-type spread.

4.5 How Are These Problems Related?

Recognition, tracking, and patch sampling make up an eclectic group of problems. In this section we propose a simple scheme to illustrate the relationships among them and among all incomplete information problems in Poisson-encounter foraging models. These three problems differ because they answer the following three questions differently: What is ambiguous? How is it ambiguous? How can the forager gain information to reduce the ambiguity? At a general level each of these questions can be answered in at least two ways.

What is ambiguous? Conventional theory takes as its starting point the moment of encounter with a patch or prey item and asks whether or in what way this item should be exploited. The decision depends on information about two things: the value of the patch or prey item at hand and the value of the alternatives, and it is assumed that the forager has complete information about both. In reality there may be ambiguity in either.

A model that treats ambiguity in the item at hand asks the question, how should a forager modify its treatment of the types it encounters when it knows that they are ambiguous? Sections 4.2 (recognition costs) and 4.4 (patch sampling) treat problems of ambiguity in the item at hand. Questions about alternatives are really questions about the environment. Models that deal with ambiguity in the alternatives ask the question, how should a forager modify its exploitation of a given item (or items) to gain information that may be useful in deciding how to exploit the items that have yet to be encountered? Our discussion of environmental tracking in section 4.3 shows that a forager may include types in its diet that are not normally included to assess whether they should now be included. Ambiguity in the item at hand and ambiguity in the alternatives are not always separable issues. For example, a forager should sometimes modify how it exploits a patch to gain information both about the patch it is in *and* about the patches it may encounter in the future. The extent to which items at hand and alternatives can be considered separately depends on the answer to the next question.

How is it ambiguous? When a forager encounters a recognized type, it is assumed to know the *prior distribution* of sub-types. If the forager knows the sub-type at the ith encounter of a foraging bout, then this may give it information about future sub-types, as it does in our model of environment tracking. On the other hand, the prior distribution of sub-types may be the same from one encounter to the next, as was the case in our discussions of recognition costs (section 4.2) and patch sampling (section 4.4).

We call these two qualitatively different types of ambiguity *sequentially dependent* and *sequentially independent* ambiguity, respectively. If the prior distribution is sequentially independent, then the ambiguities of items at hand and alternatives can be readily separated; because the item at hand can give the forager information only about itself, not about future encounters. When there are sequential dependencies, the problems of items at hand and alternatives cannot be so easily separated. The degree to which they can be separated depends on the problem of interest; for example, it may depend on whether a patch or a prey model is being considered.

How can the forager gain information to reduce the ambiguity? A forager may gain information in two ways. First, it might be able to distinguish sub-types without making any decision about whether or how to exploit the patch or prey item at hand. In other words, the forager may be able to use cues (possibly at some cost) to recognize sub-types in the way that the shore crabs (*Carcinus maenus*) studied by Elner and Hughes (1978) distinguished profitable from unprofitable sub-types of mussels by lifting them. This is the situation modeled in section 4.2 (recognition costs). Second, a forager may gain information about sub-types when exploiting them; the forager may *sample*, as we assumed in sections 4.3 (environmental tracking) and 4.4 (patch sampling). Sampling differs from *prior* recognition. When a forager uses prior recognition, it pays the price of recognition before exploiting the item. When a forager uses sampling it pays the price of recognition while exploiting the item. If the forager pays the price of recognition before the item (patch or prey) is exploited, then we can logically separate the recognition costs from foraging gains. Sampling, however, links the recognition costs to the way the item is exploited. This linkage occurs in two ways: (1) sampling depletes (although sometimes insignificantly) the value of the patch or prey item sampled and (2) the cost of sampling and the degree of recognition sampling provides both depend on the nature of the sub-types and the prior distribution of sub-types.

For example, imagine that a particular patch type consists of three sub-types. Each sub-type has at most 1 prey per patch, and the sub-types differ only in the rate (λ) at which the prey can be found: $\lambda = 0$, 1, or 2 prey per minute, according to an exponential distribution of encounter times. The three sub-types are equally abundant. If the forager hunts for 0.93 minutes without finding a prey, then it will be 90% certain that the patch is not sub-type $\lambda = 2$ prey per minute. It will have to search another one and one-quarter minutes to be 90% certain that there are no prey in the patch. The sampling forager must pay different prices (measured here in searching time) for the same amount of certainty about each alternative.

Figure 4.6 presents a classification of information problems showing that there are eight possible ways to answer the three questions posed above. This scheme is not complete, and its elements are not independent parts of the problem. The diagram simply illustrates the relationships between incomplete information problems, and it suggests new problems for study.

Branch 2 of Figure 4.6 (patch sampling) is by far the best-studied information problem in foraging theory, even though it is only one of eight information problems. This underscores the need for more theoretical and empirical work. For example, the figure shows that the patch sampling literature has concentrated on problems of sequentially independent ambiguity. These models tacitly suppose that foragers gain information only about the present patch, not about future patches.

Tracking models fit on branches 4 and 8, because items at hand and "alternatives" are both ambiguous. When the forager exploits the varying prey type (section 4.3), the item at hand is ambiguous (branch 4), but when the forager exploits the unambiguous mediocre prey types, the alternative is ambiguous (branch 8). When alternatives have sequentially independent ambiguity (branches 5 and 6), then conventional "deterministic" models apply despite the presence of ambiguities (as they did in the $q = \frac{1}{2}$ case in the tracking model of section 4.3). Here, the alternatives can be adequately characterized by their means, since the sub-types do not change predictably with time.

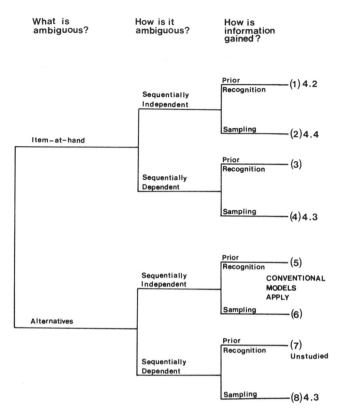

Figure 4.6 A classification of information problems. The numbers in parentheses at the end of each branch are used to distinguish one problem from another. The numbers not in parentheses are the sections of Chapter 4 that review each problem.

4.6 Summary

Following the terminology of the basic foraging models, we define a type to be a recognizable class of prey items or patches, but types may be composed of many indistinguishable sub-types. A forager's knowledge of the likelihoods of various sub-types can be represented by a prior distribution. Bayes' theorem incorporates experience and the prior distribution to yield a posterior distribution. Simple recognition cost models present opportunities to study the value of information experimentally.

Tractable models of environmental tracking can be constructed when sub-types can be readily distinguished. These models lead to three conclusions: (1) the optimal amount of tracking should represent a trade-off between the costs of sampling and the potential loss of opportunity from not tracking carefully enough; (2) tracking is generally not the best policy when the environment is unstable or when the varying prey type and its mediocre alternatives are very different in average value; and (3) although the instability of environments makes tracking necessary in the first place, information gained about unstable attributes of the environment is not as valuable as information gained about stable attributes of the environment. If the forager cannot readily distinguish sub-types, then a "comparison effect" must be added to this list. Sub-types are difficult to tell apart when they overlap, and sampling is less informative.

The marginal-value theorem overlooks the importance of patch sampling as a reason for leaving patches. Patch sampling can produce non-trivial leaving policies even if there is no physical patch depression. Nevertheless, the qualitative relationship between increasing travel time and increasing patch residence time is still expected under patch sampling models. A forager using a marginal-value rule (assessing the instantaneous rate of gain while hunting in a patch, and leaving when this rate drops to the average habitat rate of gain) may do poorly when compared with a forager adopting the optimal assessment strategy. The difference between these two policies becomes greater as the spread of patch sub-types is increased.

Incomplete information represents an eclectic set of problems. A scheme to illustrate the relationships between information problems is based on three questions: (1) what is ambiguous; (2) how is it ambiguous, and (3) how can the forager gain information to reduce the ambiguity?

5   The Economics of Choice: Trade-offs and Herbivory

5.1 Introduction

Chapters 2, 3, and 4 surveyed models with varying constraint assumptions, but all using the same currency: long-term average rate-maximizing. An obvious limitation of these models is that animals frequently face not only the problem of harvesting energy but also conflicting demands. The best feeding site may be the most dangerous, the worst place to find a mate, or the least suitable for building a nest. For a herbivore, finding plants that make up a balanced diet or ones that are not poisonous may be more important than finding enough energy. How can we use optimization models to analyze how animals might solve such trade-offs? This chapter considers this question. Here we outline an approach based on microeconomic theory and compare it with rate-maximizing; in Chapter 7 we tackle the same problem with a different theoretical tool, dynamic optimization models.

Both biologists (e.g. Covich 1971, 1972, Rapport 1971, 1980, 1981, Rapport and Turner 1977, McFarland and Houston 1981, Dawkins 1984) and psychologists (e.g. Lea 1978, 1983, Rachlin et al. 1981, Allison 1983, Staddon 1983) have recognized that the theory of economic allocation of a limited budget can be used to study animal choices. To begin, we briefly outline the relevant economic theory and review how psychologists have applied these principles before we consider how the same approach can be applied to foraging trade-offs.

5.2 Economics of Consumer Choice

Microeconomic theory uses three components to predict a consumer's choice between alternative packages of goods: *utility*, *income*, and *price* (Henderson and Quandt 1971, Mansfield 1979). Choices are assumed to *maximize* utility subject to income and price constraints. *Utility* is the "level of satisfaction" that the consumer derives from a particular package of goods. At first sight, this definition may seem vague and unrevealing to the student of foraging behavior, but the application of utility will become apparent shortly. There are several important points to note about economists' use of the term. First, economists use utility as a *descriptive* tool: they measure utility by observing what people choose, not by independent criteria. In Chapter 1 we drew attention to the distinction between this purely descriptive (*a posteriori*) approach and the *a priori* specification of utilities (or currencies) used in this book up to now. Second, economists usually only measure utility on an ordinal scale (A is better than B, not A is 2.1 times better than B). Third, economists define commodities so that the consumer prefers more to less. If the consumer prefers less of substance A, then economists consider "the commodity" to be the negative of A. Finally, economists assume that the utility derived from an additional quantum of a commodity decreases with the amount of the commodity already obtained, producing the utility function shown in Figure 5.1(A). This law of diminishing marginal utility, as it is called, has intuitive

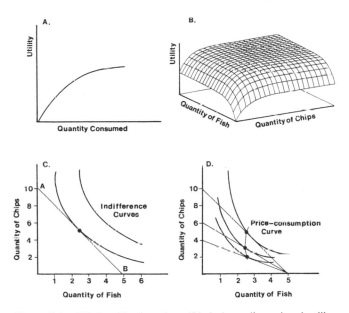

Figure 5.1 (A) A utility function. (B) A three-dimensional utility function that shows how indifference curves are isoclines of utility functions. (C) Indifference curves and a budget line (AB). (D) Effect of changing the slope of the budget line on the optimal mixture of goods. The slope of the budget line is − (Price of fish)/ (Price of chips).

appeal for commodities as diverse as computers and caviar, but we will say more about utility functions in Chapter 6.

If two commodities, for example, fish and chips (North American readers may substitute hamburgers and fries), have utility functions like the one in Figure 5.1(A), how will the utility-maximizing consumer choose between them? Figure 5.1(B) shows that many combinations of fish and chips give the same total utility (e.g. three pieces of fish and two packets of chips is equivalent in utility to two pieces of fish and three packets of chips). The consumer should be equally likely to choose either one when presented with two alternatives of the same utility; in other words the packages lie on the same *indifference curve* (Fig. 5.1[C]). A package on a higher curve will, however, always (because more is preferred to less) be preferred to one on a lower curve.

If one could build indifference curves, for example, by offering people choices between different packages and measuring preferences, one would predict that in a new choice test the individual consumer would choose a package on a high curve in preference to one on a low curve but would not discriminate between packages on the same curve. But we must also consider the constraints on choice. In economic theory income and price constrain choice. Suppose, for example, that you have only 5 dollars, and that fish costs 1 dollar a piece and chips are 50 cents a packet: you would be constrained to choose a package containing not more than 5 pieces of fish and no chips or 10 packets of chips and no fish, or any combination lying on a line such as AB in Figure 5.1(C) that joins these points together. We can now specify consumer choice: if the consumer maximizes utility subject to budget constraints, then the preferred package will be the one on the highest indifference curve that intersects the budget constraint line (Fig. 5.1[C]).

We can make many deductions about the effects of changes in income or price from this theory of choice. For our purposes the effects of changing price are most interesting, since the price of a commodity is analogous to the search costs or search time for a particular prey type (Box 5.1). Changes in a commodity's price affect the budget line's slope (Fig. 5.1[D]). If the price of chips goes up, fewer chips can be bought within the income constraint, so the budget line becomes steeper and, since the point of intersection of the budget line with the highest indifference contour has moved, the predicted fish and chips mixture changes. Figure 5.1(D) shows the effect of several different price changes, and the solid dots indicate the optimal mixture of fish and chips for the consumer at each price; the line joining the dots is the price-consumption curve.

This can be replotted in a different way, by graphing the quantity of chips consumed as a function of their price: economists call this the de-

BOX 5.1 FORAGING TIME AND THE PREY AND PATCH MODELS

This box shows, following Winterhalder (1983), how the indifference curve analysis of section 5.2 can be related to the rate-maximizing models of Chapters 2, 3, and 4. Winterhalder argues that during some time T the energy an animal acquires from foraging (E_a) must equal the energy expended (E_e), and obviously the time spent foraging (T_f) plus the time spent not foraging (T_{nf}) must equal T. The ordered pair (T_{nf}, E_a) specifies the other two variables ($T_f = T - T_{nf}$ and $E_e = E_a$).

Winterhalder treats choices of E_a and T_{nf} in the way that we have discussed choices of fish and chips. (Notice that Winterhalder studies energy acquired and non-foraging time because these are both commodities of which more is better.) Figure B5.1 shows hypothetical indifference curves for various combinations of energy acquired and non-foraging time. The curves labeled f_5 yield a higher utility than the curves labeled f_4, and so on.

When the indifference curves are nearly horizontal, as in Figure B5.1(A), then only a small increase in energy gain is required to compensate for a large reduction in non foraging time: the forager is energy-limited. When the indifference curves are nearly vertical, as in Figure B5.1(B), then a large

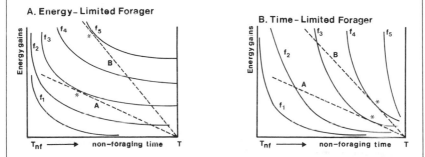

Figure B5.1   The solid lines are indifference curves that relate various combinations of energy gain (E) to non-foraging time (T_{nf}). A forager would prefer any T_{nf}—E combination on a higher curve to a combination on a lower curve: the curve marked f_5 is preferred to the curve marked f_4, and so on. The broken lines are budget lines. They show how much energy a forager can actually gain if it spends a given amount of time not foraging (that is, total time minus foraging time). The budget line marked A would occur in a habitat with a lower rate of energy gain than the budget line marked B. The optimum energy gain, marked with an asterisk, occurs at the point at which the highest possible indifference curve is tangent to the budget line. (A) shows indifference curves for an energy-limited forager, and (B) shows indifference curves for a time-limited forager.

BOX 5.1 (CONT.)

increase in energy is required to compensate for a small reduction in non-foraging time: the forager is time-limited.

The models in Chapters 2, 3, and 4 specify rates of energy intake: a given patch residence time specifies a given rate of energy intake during actual foraging. A rate plus the time spent foraging (T_f) specifies the relationship between energy acquired (E_a) and non-foraging time (T_{nf}): $E_a = R(T - T_{nf})$. This equation has slope $-R$. Its energy intercept equals RT, and its time intercept equals T, as Figure B5.1 shows. Since a higher rate can mean both more energy and more non-foraging time, rate-maximizing is preserved in Winterhalder's utility-maximizing model. However, Winterhalder's model predicts something that standard rate-maximizing cannot: how long should the forager forage? The solution is arrived at by finding the point at which the highest indifference isocline is tangent to the highest possible rate line (marked by asterisks in the figure). Notice that an energy-limited forager should forage longer, but a time-limited forager should dispense with its foraging chores more quickly. The main difficulty with Winterhalder's model is that the indifference curves are entirely hypothetical, and measuring these curves seems practically impossible. However, the model does show how, in principle, one might predict the allocation of time to foraging. Hawkes et al. (1985) have proposed a similar model.

mand curve (Fig. 5.2[A]). Demand curves make two relevant points: price elasticity and substitutability. These two related notions tell us how indispensable a commodity is, and how easily it can be replaced by an alternative. The elasticity of demand is the percent change in demand arising from a percent change in price (note, incidentally, that because elasticity is a relative measure it is not simply the slope of the demand curve). A commodity with inelastic demand is one for which the percent change in consumption is less than the percent change in price. Such commodities are likely to be ones that are essential for survival; if an animal needs a certain amount of food per day and food gets scarcer (price goes up), the animal might be expected to spend more time looking for food instead of decreasing its hunting time and doing something else (see Box 5.1). In contrast, commodities with elastic demand are those for which demand goes down proportionately faster than the increase in price. A possible animal example is sexual display (Hogan et al. 1970; see McFarland and Houston 1981).

Figure 5.2   (A) A demand curve. (B) Demand for one commodity as a function of the price of another. Two cases are shown: substitutable and complementary resources.

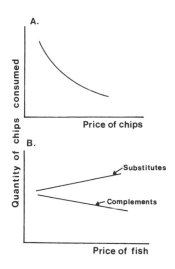

The elasticity of demand for a commodity may vary according to circumstances such as total income, preference, and the availability of alternatives. If income and preference remain constant, then available alternatives determine elasticity: for example, if you think that chips are an adequate alternative to fish, when the price of fish goes up you would be likely to spend less on fish (possibly even nothing) and buy chips instead. Fish would show elasticity of demand and fish and chips would be said to be substitutes for one another. Figure 5.2(B) shows this point graphically by indicating that as the price of fish goes up, the quantity of chips consumed increases. The same graph also plots the opposite case, in which resources or commodities are complementary: an increase in the price of one causes a decrease in consumption of the other. Croquet mallets and balls are an example. Since these two commodities are used in a set, an increase in the price of croquet balls will cause people to buy fewer croquet sets and therefore will cause a decrease in the demand for mallets. (Note, however, that when used for other purposes, for example as missiles to be hurled at the neighbor's dog, mallets and balls could become substitutes.)

A final point to note about substitutability is that substitutable commodities have indifference curves that are nearly (depending on how substitutable they are) linear, but complementary items have strongly curved indifference contours. This fact crops up in the discussion below.

How is this economic theorizing relevant to foraging trade-offs? In principle one could design an experiment to find out which combinations of commodities foraging animals find equally valuable (e.g. mixtures of two types of food, mixtures of food and safety from predators) and then use the economic approach we have outlined to predict choice under known time-budget constraints. Similarly, the substitutability of different food types or activities could be inferred by observing how changes in the availability of one commodity affect demand for the other. In addition to these descriptive uses, *a priori* principles can be used to specify the shape of the utility function and hence the indifference contours; the concepts of elasticity and substitutability can be used regardless of how the indifference contours are found (see Box 5.1 and Chapter 6 for examples).

5.3 Economic Choice and Animal Psychology

This section discusses how these ideas have been used in animal psychology. Our commentary is highly selective, and more extensive reviews can be found in Lea (1978, 1983), Rachlin et al. (1981), Allison (1983), and Staddon (1983). Criticisms of the economic approach as it is applied in psychology may be found in the discussion following Rachlin et al.'s paper and in Vaughan and Herrnstein (1986).

Psychologists have used economic theory both as a model for animal choice and to model human economic systems, such as the labor market, using animals. In both cases psychologists usually present animals with food or some other consumable commodity in an operant experiment in which the animal must work to obtain the commodity, and in which both the amount obtained and the budgetary constraints can be precisely controlled. Psychologists use this approach to account for general behavior allocation patterns (e.g. percentage of choices allocated to each alternative), rather than the moment-to-moment decision rules used by the animal in making its choices (see Chapter 8).

Rachlin et al.'s (1976) classic study of the substitutability of two commodities provides a good illustration of how the economics of choice can be studied. These experiments also show how one might study the diet choice of animals with suspected or known complementary dietary components. Rachlin et al. trained rats to press two levers: the rats got root beer from one lever and Tom Collins mix from the other. A ratio schedule (in a *ratio schedule* a fixed number of responses on the appropriate lever is required for each commodity) determined the delivery of fixed small amounts of liquid. Rachlin et al. limited the total number of responses per experimental session (an income constraint) and the number of responses (prices) required to obtain each commodity. This gave the experimenters precise control of the budget line. They first determined the allocation of responses to the root beer and Tom Collins levers when prices were equal. They then increased the price of one, say root beer. What should the rats do? If the two items are substitutable, the rats should spend more of their presses on the cheaper alternative and decrease expenditure on the expensive one, but if the resources are not substitutes, there should be little change in the relative amounts consumed. Figure 5.3 explains this behavior graphically. Figure 5.4 shows the outcome of both the soft drinks experiment and another one involving the choice between food and water. The two liquids turned out to be nearly substitutable (note that if they were perfect substitutes and if the rats were maximizing utility, then the rats would have chosen only the cheaper alternative, as they do when the two alternatives are identical—Herrnstein and Loveland 1975). In contrast, when the two substances are food and water, changing the price has little effect on the relative amounts consumed; as an ecologist would expect, food and water are not substitutes. The general point here is that one can draw inferences about substitutability and the shape of the indifference curves by changing the commodity prices (availabilities) in simple choice experiments.

A second illustration of the use of economic principles in animal psychology concerns how rats trade off work and leisure (Rachlin et al. 1981,

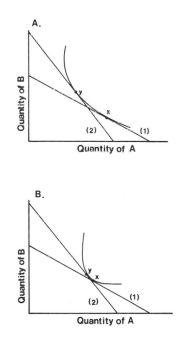

Figure 5.3 (A) When two resources are near-perfect substitutes, the indifference curves are fairly shallow, and an increase in the relative price of one of them—an increase in A's price is indicated by the steeper slope of line (2)—causes a big change in the mixture consumed (from x to y). (B) When the commodities are complements, the indifference curve is nearly a right angle, and a change in the slope of the budget line produces only a small change in the mixture consumed (x to y).

Figure 5.4   (A) When Rachlin et al. changed the relative price of two kinds of soft drinks (changing the budget line from AB to CD), their rats shifted their preferences by amount X. (B) For food and water, however, the preferred package only changed by amount Y. Soft drinks were more substitutable than food and water.

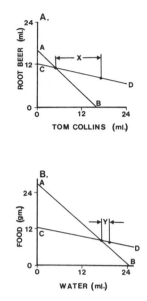

Allison 1983). The concept of leisure may seem artificial to the behavioral ecologist, but it could be replaced with any other activity for purposes of analyzing a trade-off problem (see Box 5.1). The experimenters placed rats (or pigeons) in a Skinner box for a fixed time each day, and they trained the animals to work for food on a ratio schedule. The experimenters varied the ratio of responses required to obtain rewards, and they measured the rate at which the rats worked during each session. Figure 5.5(A) shows the result: as the ratio gets smaller (reward rate increases), the animals at first increase their work rate, but at high reward rates they slow down again: the maximum work rate occurs at intermediate reward rates. Microeconomic analyses of the effect of wages on labor supply predict just this relationship. Figure 5.5(B) summarizes this argument (see Allison 1983 for an alternative interpretation).

In these examples, psychologists inferred the shape of the indifference contours from the animal's choices. Maynard Smith (1978) calls this the "reverse optimality" approach, and he criticizes it for doing nothing more than describing the data. However, a description of the indifference contours resulting from one set of observations can be used predictively in another situation. If the model survives this test of "trans-situationality," then it may represent some general properties of the animal's preferences. Although the approach seems suitable for the study of trade-offs between

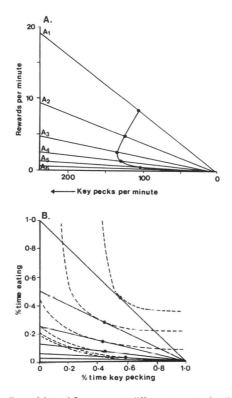

Figure 5.5 (A) The lines A1 to A6 represent different constraint (or budget) lines set by the number of key pecks required to get a single reward. For example, on line A1 a rate of 200 pecks per minute will yield about 17 rewards per minute (ratio $= \frac{200}{17}$), but on line A6, 200 pecks per minute will yield only 1 reward ($\frac{200}{1}$). When allowed to work for a fixed time each day, on different ratios, pigeons chose how much to work and eat (the rates shown by the solid dots). Work rate was higher at intermediate ratios. (B) The positions of the solid dots in (A) could be described by assuming that the animals maximize utility subject to varying constraints (the different ratio schedules). In (B) hypothetical indifference curves of combinations of feeding rate (I) and leisure (L, negative of work rate) are shown. On each ratio schedule the animal chooses the highest indifference curve it can obtain: the points of intersection of the curves and the constraint lines trace out the pattern shown by the animals.

activities such as foraging and vigilance, or foraging and territorial defense, it has not been widely used in behavioral ecology, an exception being the study of Kacelnik et al. (1981) on the trade-off between foraging and territorial defense in great tits. In the next section we discuss how behavioral ecologists have studied foraging trade-offs.

5.4 Studies of Trade-offs: Birds Are Tame in Winter

There is a danger that qualitative studies of trade-offs may not progress beyond conclusions such as the one in the heading of this section. We might view the boldness of normally shy birds that come to feeders in cold winters as evidence of a trade-off between feeding and predator avoidance, but this conclusion is a bit limp. Apart from the descriptive utility approach described in sections 5.2 and 5.3, behavioral ecologists have tackled the trade-off problem quantitatively in two ways: maximizing energy gains subject to time-budget constraints imposed by competing activities and using survival as a common currency. Both of these are *a priori* approaches.

Time budgets. The analysis of wagtail territorial defenses by Davies and Houston (1981) provides an excellent example of this quantitative approach. The pied wagtail (*Motacilla alba*) generally excludes all other wagtails from its winter feeding territory, but occasionally it shares its territory with another wagtail (a "satellite"). Davies and Houston consider the economics of tolerance versus exclusion in terms of time. Excluding the satellite increases defense time (because the satellite helps to keep others out of the territory), but tolerating the satellite decreases the owner's feeding rate (because it competes with the satellite for food). Thus both defense and feeding costs can be expressed as losses or gains in time and incorporated into the animal's feeding rate. Davies and Houston used field estimates of cost to calculate whether there would be a net increase or decrease in feeding rate as a result of tolerating a satellite for each of several days. The hypothesis that the owner tolerates a satellite only when this increases its feeding rate accounted for the behavior on 35 out of 41 days. Davies and Houston reduced the trade-off problem to maximizing feeding rate subject to various time-budget constraints associated with territorial defense. This approach implicitly assumes that there are no costs other than lost feeding time associated with different activities; for example, the risk of injury when chasing a rival must be the same as the risk during feeding.

Survival rate. Caraco et al. (1980a) used the currency of over-winter survival to study the effects of group size on the allocation of time to different activities: feeding, scanning for predators, and fighting. They were not, however, able to determine directly the effects on survival of variation in time allocation, so their model was based on guesswork and could not be used to make more than qualitative arguments. Chapter 6 discusses in some detail the use of survival as a currency. Box 5.2 presents a qualitative

BOX 5.2 RATE-MAXIMIZING VERSUS PREDATOR AVOIDANCE

Lima et al. (1985) have studied a simple situation in which rate-maximizing conflicts with predator avoidance. Squirrels may have to leave the safety of cover to reach food patches, and when patches consist of many prey items squirrels must "decide" whether to eat the prey out in the open or carry them back to cover, where they can be eaten in relative safety.

Suppose that patches contain N items. The forager decides how many items to carry back to cover (N_c), and the travel time from cover to the patch is τ. Assuming that the last item is always carried back to cover (i.e. $1 \leq N_c \leq N$), there are N_c round trips from cover to the patch, and total exploitation time is

$$2\tau N_c + Nh,$$

where h is handling time. Thus the rate of energy gain is

$$\frac{Ne}{2\tau N_c + Nh},$$

where e is the mean energy value per prey item. Obviously, increases in N_c always decrease the rate of intake, so if the forager based its decision solely on rate-maximizing it would only carry the last prey item back to cover. However, Lima et al. suppose that being "in cover" is safe, but that there is a constant probability of predation (α) when the forager is exposed. The forager is out in the open for

$$2\tau N_c + (N - N_c)h$$

time units. Thus the probability of surviving a patch exploitation bout is

$$\pi(N_c) = e^{-\alpha[2\tau N_c + (N - N_c)h]}.$$

To study how N_c affects the probability of survival, consider the case in which the difference $\pi(N_c + 1) - \pi(N_c)$ is greater than zero. A little algebra shows that the probability of survival increases with the number of items carried if the handling time is greater than the round trip time ($h > 2\tau$); the probability of survival decreases with the number of items carried if the round trip time is greater than the handling time: rate-maximizing and predator avoidance are *not* in conflict here. Predator avoidance only conflicts with rate-maximizing when the handling time is greater than the round trip time.

If the forager's behavior represents a compromise between these two goals, then we would expect that (1) more items should be carried back to cover when the distance to cover is short and (2) if there is an increasing relationship between prey size and handling time, then more items should be carried back when prey sizes are large. In an empirical test with grey squirrels (*Sciurus carolinensis*) Lima et al. (1985) found that squirrel behavior was consistent with these qualitative predictions (see also Lima 1985b).

example that considers the conflict between feeding rate and exposure to predators.

5.5 Nutrients and Diet Choice by Herbivores

Many ecologists would agree that herbivore diets, especially generalist herbivore diets, are much more complex than the diets of other consumers: rate-maximizing might (it is argued) explain carnivore diets, which are made up of prey with approximately the right balance of nutrients, but herbivores often feed on abundant low-quality prey and face the problem of selecting a balanced diet and not just maximizing the rate of energy gain (Crawley 1983). Furthermore, many plants contain toxic substances, adding to the herbivore's dietary woes. This view of complex qualitative effects is, in part, a justified reaction to the older view that herbivores cannot be food-limited because the world is green (Hairston et al. 1960); nevertheless, we will argue in this section that simple rate-maximizing models may still be valuable in interpreting herbivory, and that this emphasis on food quality for herbivores has stymied attempts to find out just how complex herbivore diet selection is. Because we cannot review the vast literature on (both insect and vertebrate) herbivory, we refer the reader to Barker et al. (1977), Harborne (1978, 1982), Rosenthal and Janzen (1979), Morley (1981), Ahmad (1983), Crawley (1983) and Denno and McClure (1983). What we present here is a view of herbivory from the perspective of foraging theory.

SOME SPECIAL FEATURES OF HERBIVORES

Given the great diversity of herbivore life styles, it may seem foolhardy to attempt generalizations, but on the whole herbivores tend to differ from the "typical predators" represented in the models of Chapter 2 and 3 in the following ways: (1) they spend little time searching for food and much time ingesting and digesting, so their time budgets differ from those of predators that consume animal matter; (2) a herbivore's food seldom occurs as neatly packaged prey items, so the concept of "encounters" may be difficult to apply; and (3) some small herbivores consume only one food item (plant) during their lifetime (or during their larval "lifetime"): for these animals, foraging decisions must be restricted to problems such as the choice of different places to feed within a plant rather than the choice of plants. These differences may mean that the decisions studied in models of herbivore foraging cannot be the same as those for "typical predators." For example, "prey choice" may be formulated as a decision about diet composition (Belovsky 1978, 1981, 1984) rather than about acceptance or rejection upon encounter (Chapters 1 and 2).

It follows from the third point above that small (usually insect) and large (usually vertebrate) herbivores may face different decisions. In the following discussion we primarily address the problems of large, generalist herbivores, but we also comment on the problems of smaller, specialist herbivores such as insect larvae.

THE DIETS OF LARGE HERBIVORES: THREE VIEWS

A survey of the literature reveals three main views about the nature of decision, constraint, and currency for herbivore diets: rate-maximizing subject to nutrient constraints, selecting complementary nutrients, and avoiding toxins.

RATE- OR AMOUNT-MAXIMIZING SUBJECT TO CONSTRAINTS. According to this view, herbivores select food items that maximize the rate of energy intake (or energy gains over a fixed time—Belovsky 1978) or nitrogen (Owen-Smith and Novellie 1982), or that maximize digestion rate (Westoby 1974, Sorensen 1984) if digestion is slower than ingestion. This view also recognizes that requirements for one or a small number of essential nutrients such as sodium (Belovsky 1978) or avoidance of poisons may constrain rate-maximizing. The following general observations about large herbivores and their food lend credence to the rate-maximizing view.

1. Detailed studies of plant chemistry often show that the concentrations and availabilities of different nutrients in plants species or parts are correlated: parts with high digestibility also tend to be high in protein (Arnold 1981). For example, Glander (1981) found that howler monkeys (*Allouatta palliata*) ate leaves with relatively high concentrations of all amino acids, and that young leaves were richer in all amino acids than older leaves, probably a general pattern for most plants. Thus selection of plant parts to maximize intake of one nutrient may frequently maximize intake of many or all nutrients at once.

2. Herbivores do not seem to be able to detect specific nutrients, apart from sodium and water, in different dietary components (see below). This makes complicated models, in which diet choice interacts with nutrient limitations, less plausible.

3. Many herbivores, including ruminants and aphids, can synthesize most amino acids and many other essential nutrients (Owen Smith and Novellie 1982, Crawley 1983) and may not require a complex diet to meet their requirements.

The effects of factors such as fiber, silica, and tannins, which reduce the digestibility of plant material (Harborne 1982), can be readily incorporated into a rate-maximizing model by including the constraints of stomach capacity, passage rate, or both. More specific chemical poisons such as alkaloids can be thought of as anti-nutrients that limit the total intake of any one dietary component.

In section 3.6 we introduced a method, mathematical programming, for calculating the rate-maximizing diet subject to nutrient or toxin constraints. To recap briefly, suppose there are two nutrients x and y and the animal's hypothesized goal is to maximize intake of x subject to the constraint of meeting a fixed requirement of y per unit time (or per stomach load if digestion is the rate-limiting step). If the proportions of x and y are known for different prey types such as P and Q, equations can be written that specify the amounts of x and y obtained as a function of the proportions of P and Q in the diet. We can use linear programming to find the proportions of P and Q that maximize the intake of x and satisfy the constraint in y. Section 3.6 and Box 3.5 provide further details and limitations.

Belovsky (1978) used this approach in his work on the diet of moose (*Alces alces;* see Box 5.3). He identified sodium as a likely nutrient constraint. Sodium is a good general "constraint" candidate for three reasons. First, vertebrates require large amounts of sodium, since they must replace the sodium lost each day in urine. Second, sodium is also the only nutrient other than water for which a "specific hunger" has been demonstrated (see below). Third, sodium is often scarce in plant food. Belovsky also hypothesized that the moose maximizes daily energy intake subject to a sodium constraint. The moose can choose any mixture of high-sodium, low-energy aquatic plants and low-sodium, high-energy terrestrial plants. Belovsky estimated the daily sodium and energy requirements of the moose and the constraint imposed by rumen size, and he showed that the proportion of aquatic plants in the diet was remarkably close to that predicted by the energy-maximizing linear programming model. Although this simple approach worked well in accounting for the use of broad classes of vegetation (i.e. aquatic versus terrestrial plants), linear programming failed to predict the moose's choice of individual species within the terrestrial habitat (Belovsky 1981), possibly because of the role played by toxins. Recently, Belovsky has extended this analysis to the diets of three generalist herbivores (Belovsky 1984, Box 5.3).

SELECTING COMPLEMENTARY NUTRIENTS. Although much work on herbivory implicitly assumes that herbivores consume plants that contain complementary nutrients, only Covich (1972) and Rapport (1980, 1981)

BOX 5.3 WHAT SHOULD A HERBIVORE EAT?

A herbivore's diet may be constrained by one or more of the following: toxin intake, nutrient intake, digestion rates, ingestion rates, and limits on daily feeding time. Trying to predict a herbivore's diet while simultaneously taking all these factors into account may seem a hopelessly difficult task. However, in some circumstances the solution is not so difficult.

Specifically, the mathematical techniques called *linear programming* maximize a currency subject to many constraints, but both the currency and the constraints must be linear functions of the decision variables. Belovsky (1978, 1981, 1984) has used linear programming to find the energy-maximizing diets of several vertebrate herbivores. We present a modified version of Belovsky's well-known model of moose (*Alces alces*) diets.

We pointed out in the text that sodium is a nutrient that is likely to affect diet choice: animals need a lot of it, and they can probably tell when they do not have enough (a specific hunger for sodium has been demonstrated in some vertebrates). However, the moose's sodium woes are even more severe than usual. Aquatic plants are the only significant source of sodium in moose habitats (e.g. Isle Royale, Michigan), and aquatic plants are not available in the winter because they are under ice. The moose must therefore get a whole year's supply of sodium during the summer. Yet another complication is that aquatic plants are bulkier (they take up more room in the gut) and supply less energy than terrestrial plants.

How much of the moose's summer diet should be sodium-rich aquatic plants, and how much should be energy-rich terrestrial plants? Linear programming can answer this question. Let A and T be decision variables: A = grams (dry weight) of aquatic plants eaten per day, and T = grams (dry weight) of terrestrial plants eaten per day. The currency (energy maximization) is maximize

$$3.8A + 4.25T,$$

because aquatic plants give 3.8 kilocalories per gram dry weight, and terrestrial plants give 4.25 kilocalories per gram dry weight.

CONSTRAINTS
Maintenance Requirement

$$14,000 \leq 3.8A + 4.25T;$$

an average moose must have 14,000 kilocalories per day simply to survive. *Digestive Limitaton*

$$32,900 \geq 20A + 4.04T;$$

BOX 5.3 (CONT.)

an average moose can process only 32,900 grams of *wet* food per day. Aquatic plants have 20 grams of wet weight for every dry gram, and terrestrial plants have 4.04 grams of wet weight for every dry gram.

Sodium Requirement

$$2.57 \le 0.003A;$$

an average moose needs 2.57 grams of sodium per summer day to make up its yearly requirement. Aquatic plants contain 0.003 grams of sodium per gram dry weight; terrestrial plants contain insignificant amounts of sodium.

Condensing all these expressions, we can state the problem in this way:

Maximize	$3.8A + 4.25T$,
subject to	$14{,}000 \le 3.8A + 4.25T$
	$32{,}900 \ge 20A + 4.04T$
	$2.57 \le 0.003A.$

We solve the three constraint equations for A and plot the resulting lines as shown in Figure B5.3. The triangular hatched region in this figure contains all the (T, A) points that satisfy the constraints.

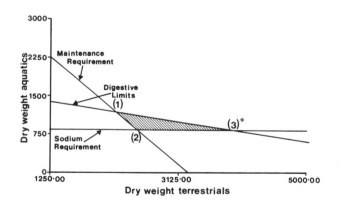

Figure B5.3 The hatched region shows all (T, A) points that satisfy the constraints on moose diet selection, where T is the dry weight of terrestrial plants consumed per day and A is the dry weight of aquatic plants consumed per day. Because the currency and the constraints are linear functions of T and A, the optimum diet composition must lie at one of the vertices marked (1), (2), and (3). Here, the energy-maximizing choice is at vertex (3).

BOX 5.3 (CONT.)

There are still many possible (T, A) pairs within the hatched region, so it may seem that we have not gotten too far. However, when the currency and the constraints are linear functions of the decision variables, the maximum currency occurs at one of the feasible region's vertices—labeled (1), (2), and (3) in Figure B5.3. Each vertex can be found by solving two simultaneous equations; here, vertex (1) is (2223 grams/day, 1198 grams/day), vertex (2) is (2527 grams/day, 857 grams/day), and vertex (3) is (3920 grams/day, 857 grams/day)—and these vertices can simply be plugged into the currency function to find which one gives the most energy per day. Vertices (1) and (2) both give 14,000 kilocalories/day, since they lie on the line specified by the maintenance requirement, but vertex (3) gives much more, 19,918 kilocalories/day. Vertex (3) is the energy-maximizing choice.

We could also use the techniques discussed in section 5.2 to solve for the energy-maximizing diet. The indifference curves between aquatic and terrestrial plants are parallel straight lines with the same slope as the maintenance constraint line in Figure B5.3. Obviously, the highest indifference line passes through vertex (3).

Belovsky's model predicts that an average moose's summer diet should be about 18% aquatic plants $[0.18 = 857/(857 + 3920)]$. Astonishingly, observations show that, on average, moose choose diets containing about 18% aquatic plants. Belovksy has found similarly staggering agreements with linear programming in the choice between grasses and forbs of the *Microtus pennsylvanicus* and in the choice between herbs and shrubs of the kudu (*Tragelaphus strepsiceros*).

Despite these instances of close agreement, linear programming has serious limitations. One problem is that neither linear programming itself nor its advocates in foraging theory can give *a priori* rules for classifying plant foods. For example, although linear programming predicts the mix of aquatic and terrestrial plants in the moose's diet, it failed to predict the moose's choice of plant species within the terrestrial habitat (Belovsky 1981). It is also important to remember that linear programming is not *a* model of diet choice, but rather a huge family of models of diet choice: there are many ways to specify the constraints on diet choice even when studying a given species of forager.

Attempts to compare the diets predicted by linear programming and the "diets" predicted by the basic prey model outlined in Chapter 2 have run afoul of these limitations (Belovsky 1984). The basic prey model cannot predict the proportion of aquatic and terrestrial plants in a moose's diet, because applying the prey model assumes that the moose searches for all prey types simultaneously: only a dimwitted moose would search for terrestrial plants in a pond.

have analyzed diets using the concept of complementarity as defined in section 5.2. According to Rapport's view, herbivores are best seen not as maximizing the rate of intake subject to constraints but as choosing between packages consisting of complementary mixtures of dietary components. The complements could be nutrients or inhibitory poisons (Levander and Morris 1970, Freeland and Janzen 1974).

To grasp more readily the difference between the "constraints" and "complements" views of nutrients, consider the case of two nutrients. The constraints view implies that the utility function for one nutrient is a step function (above the step the requirement is met, below it is not), although for the other nutrient it is a monotonic function like that in Figure 5.1(A). The resulting indifference contours are straight lines (or nearly straight lines; see Fig. 5.3[A]), since the two resources are substitutable, and the constraint sets a lower boundary on the contours (Fig. 5.5). In contrast, the indifference contours for complementary resources will be like that shown in Figure 5.3(B): the contours for perfect complements would be right angles, but assuming some substitutability, they will be curves.

The effect of changes in prey abundance on diet composition according to the two models can be seen by superimposing budget lines of different slopes on the indifference contours, as explained in Figure 5.3. When the resources are imperfect complements, a change in the abundance of either one will alter the diet composition (an increase in abundance of A or B will lead to an increase in its representation in the diet). The constraints model makes a different prediction, more akin to that of conventional diet models. Suppose for example, as in Belovsky's study, that there is a more profitable prey type and a less profitable type, and the less profitable type contains an essential nutrient: changes in abundance of the less profitable type should not influence its inclusion in the diet, but the more profitable type should be taken in proportion to its availability (see Box 3.5).

What is the evidence that diet selection in herbivores should be viewed as selection for complementary nutrients rather than as rate- or amount-maximizing subject to constraints? Rapport (1981) supports the "complementarity view" by showing that protozoa (and, it turns out, many other animals, including insectivorous birds—Krebs and Avery 1984) grow better on mixed diets of natural prey than on pure diets. But this observation could equally well be interpreted as the effect of nutrient constraints. Two studies have more directly applied economic notions of complementarity to diet choice.

In a pioneering study Covich (1972) studied *Peromyscus* seed preferences in the same way that Rachlin et al. (1976) studied rats choosing between soft drinks. Covich established the proportion of *Cucurbita* and *Helianthus* seeds consumed by *Peromyscus* under *ad libitum* conditions and then altered the availability of each seed type in turn. The results suggested that the two seed types were partially substitutable (as in the "root beer–Tom Collins" experiment), but the results were ambiguous because Covich had only limited control over the budget line, and because training effects may have affected long-term food preferences (Partridge 1981). Importantly, Covich's work shows how economic theory can be applied to the study of diets. The only other published study to date that used indifference-curve analysis is Rapport's (1981) study of the protozoan *Stentor* feeding on pairwise mixtures of *Euglena*, *Tetrahymena*, and *Chlamydomonas*. Rapport demonstrated three main points: (1) *Stentor* always consumed a mixed diet when more than one prey was available; (2) the preference for a particular prey species increased as a function of its abundance, as might be expected if the indifference curves were of the form shown in Figure 5.3(B) for complementary resources; and (3) there was some indication that *Stentor* reproduced faster on mixtures of algal and non-algal prey than on pure diets, although the results were variable. These results do not distinguish between the constraints model and the complementarity model.

AVOIDING TOXINS. Freeland and Janzen (1974), Rosenthal and Janzen (1979), and Harborne (1982) support the view that plant toxins determine herbivore diet choice. Feeny (1975) and Rhoades and Cates (1976) were the first to distinguish two types of chemical defense: some, such as tannins, lower the digestibility of plant proteins for all herbivores when defensive chemicals are sufficiently concentrated; other, more poisonous toxins, such as cyanogens and cardenoides, work at low doses and are often overcome by specialist herbivores. As we have already mentioned, the "tannin-like" toxins can be treated within conventional rate- or amount-maximizing models by considering assimilated rather than gross intake (which is what should be done in any case). The more poisonous toxins may limit the range of prey types available to a herbivore (this is the thrust of much of the insect herbivory literature), just as an insect's defenses may exclude it from an insectivore's diet. The more poisonous toxins do not usually affect specialist insect herbivores when they feed on their usual host plant (Blau et al. 1978), even though these specialists may be unable to tolerate low doses of toxins from other plant species; thus the limiting effect of the second class of toxins on diet breadth may be severe. For large, generalist herbivores the picture is less clear: some authors think that the more poisonous toxins strongly influence diet choice (e.g. Bryant and Kuropat 1980), and in other studies nutrient quality and presence of tannins seems to be more important (e.g. Arnold 1981, Glander 1981).

NUTRITIONAL WISDOM IN LARGE HERBIVORES

The extent to which foragers can fine-tune their nutrient intake may depend on the existence of "nutritional wisdom," an idea Richter (1943) developed in his classic studies of rats. When presented with a variety of mineral sources, yeast, casein, dextrose, olive oil, and so on, rats managed to select a balanced diet. Further, when deprived of one essential nutrient, they could compensate by adding a new item to the diet (e.g. when yeast was removed, the rats started to eat their own feces, an alternative source of vitamins of the B complex).

Richter tried to explain this behavior by proposing "specific hungers" for each of the different essential nutrients, but later work has shown, at least for rats, that this is not the mechanism of nutritional wisdom. Salt and water (and possibly sugar) are the only dietary components for which rats can both detect a deficiency in their body tissue and select new diets to compensate for the deficiency. Thiamine-deficient rats, for example, can learn to avoid diets that do not contain thiamine, but they do not specifically select thiamine-rich diets from an array of novel options: they find the right one by trial and error. Animals compensate for deficiencies of other essential nutrients by learning to avoid diets with adverse consequences (including those arising from nutrient deficiency) rather than through specific appetites for the missing nutrient (Rozin 1976, 1977 for reviews). For a rat, an essential part of this process is its habit of eating more or less discrete meals consisting of only one dietary item at a time— this allows the rat to experience the consequences of each food type. If other vertebrates have the same mechanism of nutritional wisdom, then we can ask how generalist herbivores might fare in balancing their diet by seeking out particular nutrients.

Zahorik and Houpt (1977, 1981) point out that there are many theoretical difficulties inherent in applying avoidance learning to herbivores. First, meals are not discrete entities, especially for herbivores with multiple stomachs. Second, meals rarely contain a single food type. Third, adverse consequences of eating often develop over a long period, and the effects of many hours or days of eating might be confounded. Finally, many plants have some adverse consequences (e.g. they contain tannins) but are an essential component of a herbivore's diet. Experimental evidence supports the view that vertebrate generalist herbivores are limited in their ability to select food using avoidance learning: sheep and donkeys cannot avoid food that has adverse consequences if the delay is long or if they eat a mixed meal (Zahorik and Houpt 1981). This suggests that herbivores are unlikely to possess sophisticated nutritional wisdom (Arnold 1981). Zahorik and Houpt (1977) propose that a rule such as "eat young growing shoots" may provide herbivores with a perfectly adequate diet: digestible and rich in protein and calories. These limited mechanisms of food selection make the rate- or amount-maximizing view of herbivory more plausible than the complementary nutrients view.

MAINLY ABOUT INSECTS: NITROGEN, TRADE-OFFS, PATCHES, AND OVIPOSITION

So far we have commented only on large generalist herbivores. In this section we discuss briefly insect specialists.

NITROGEN. Herbivorous insect larvae, which need materials for growth, may differ in their diet selection from generalist herbivores, which may be more likely to select energy-rich foods. Many authors have suggested (notably Scriber and Slansky 1981) that nitrogen is more important to these larval herbivores than energy is. Building and testing alternative nitrogen and energy-maximizing models would not be difficult, but there has been little work in this direction. Although nitrogen concentration (McNeill and Southwood 1978) and toxins (McKey 1979) influence the selection of plant parts by insect herbivores, avoidance of predators also plays an important role (Thompson 1982).

TRADE-OFFS. Insects may adopt costly behaviors such as migrating from daytime resting sites to remote feeding places at night and then back again at dawn (e.g. Heinrich 1979). Presumably this migration allows the insect to rest in a safe refuge away from the telltale signs of leaf damage at its feeding site which might attract predators. As such, it is similar to the trade-offs discussed in section 5.4. The insect can be thought of as trading off patch quality and predation risk (Schultz 1983). The same theoretical approaches outlined in section 5.4 could be applied, but they do not appear to have been to date.

PATCHES. Insect (and other) herbivores do not encounter and eat discrete prey items of the kind that the basic prey model imagines: instead of eating all-or-nothing when encountering a leaf, a herbivore is more likely to eat part of a leaf. But this does not mean that foraging theory cannot be applied to herbivore diet choice. It simply means that patch-use models are more appropriate than the basic prey model (Chapter 2). For example, a patch-use model might predict that lower-quality plant parts should be acceptable when the inter-plant travel time is longer. Despite the explosion of empirical work on how travel times effect the patch-use behavior of non-herbivores (Chapter 9), there has been little interest in this problem among students of herbivory. One exception is Parker's (1984) work, which shows that grasshoppers (like non-herbivores) are more persistent in patches when the inter-patch travel time is long. One difficulty is that plants can be patchy in subtle ways: leaf quality often varies within a single plant (e.g. Schultz 1983, who argues that even within a tree leaves vary in nutritional quality, physical exposure, and danger from predators) and between plants within a species (e.g. Denno 1983). Insect herbivores are known to respond to this kind of patchiness in their prey (Stanton 1982, Denno 1983).

OVIPOSITION. Several authors have compared the oviposition decisions of female insects with phytophagous offspring to foraging decisions (Jaenike 1978, Mitchell 1981, Stanton 1982, Rausher 1983a, 1983b). Egg-laying butterflies appear to make "accept or reject" decisions upon encounter with individual host plants, and at least some studies have found that these butterflies lay eggs preferentially on species or individual plants within a species that are better than average for larval growth and survival (Rausher and Papaj 1983). Furthermore, the probability that a female will lay eggs on a given plant sometimes decreases as the density of host plants increases (Rausher 1983b), and this effect has been compared to the basic prey model's prediction that selectivity should increase with increases in the abundance (encounter rate) of highly ranked prey. However, the comparison has not been fully worked out. To do so will require specifying an appropriate currency (rate of egg-laying, rate of larval survival, lifetime reproductive success?) and constraints (what is the analogue of handling time?). Jaenike's (1978) model of oviposition is superficially similar to the diet model and predicts that selectivity should increase with increased host-plant abundance, but the prediction does not arise from the principle of lost opportunity as it does in the prey model (Chapter 2). Instead, Jaenike assumes that the female has limited egg-carrying capacity and therefore must lay eggs to make room for new ones. Jaenike hypothesizes that the female maximizes survival per egg laid, subject to the constraint that the eggs must be laid within a certain time. As this time approaches, the female may dump her remaining eggs on less suitable host plants, and this means that egg-dumping is more likely when preferred hosts are scarce. It is not clear whether Jaenike's model uses an appropriate currency; at least it would be valuable to compare his model with models that use alternative currencies, such as maximizing lifetime reproductive success.

5.6 Summary

This chapter considers analyses of the trade-offs between foraging and other activities, such as scanning for predators. Economic models based on maximizing utility subject to budget constraints might be used to analyze trade-offs. Psychologists have successfully applied this approach (using *a posteriori* optimization), but few behavioral ecologists have used it. Alternative, *a priori* approaches to the problem of trade-offs include treating the trade-offs as time constraints and using energy maximization as the currency, and converting the consequences of each behavior to probabilities of survival. Chapter 6 discusses survival models in more detail.

Although many features distinguish vertebrate herbivores from other predators, we suggest that diet selection in generalist herbivores may be profitably studied by assuming rate- or amount-maximizing subject to suitable constraints. The literature on nutritional wisdom does not reveal sophisticated mechanisms for selecting a balanced diet. The idea that herbivores (or other predators) select a diet of complementary nutrients is frequently implied but seldom tested. The economic concept of complementarity shows how such rigorous tests might be performed. For insect herbivores foraging may best be viewed as a patch-use problem. Oviposition decisions of adults with herbivorous larvae are only superficially similar to foraging decisions.

6 Risk-Sensitive Foraging

6.1 Introduction

When we referred in Chapter 5 to "making a choice" (choosing a package of goods), we also assumed that if a forager chooses a habitat with 100 prey per square meter, then this is exactly what the forager gets: choices yield completely certain and predictable results. However, foraging behavior probably cannot specify "a package" without error; a more reasonable claim is that the forager chooses between probability distributions. The study of risk deals with this kind of choice. For example, the forager may choose a habitat in which it sometimes gets 100 prey per square meter but sometimes gets 20 prey per square meter. This chapter asks how foragers can choose between probability distributions to maximize pay-off, and how these choices can be described. But first we make a brief comment on terminology: we use the word *risk* only to mean probabilistic variation. This definition is distinct from the alternative meaning, "risk of predation," commonly used in the foraging literature. To avoid confusion, we call this second kind of risk *danger*.

6.2 Risk and Utility

Chapter 5 introduced the economist's concept of utility and showed how it can describe choice even when utility can be measured only on an ordinal scale (e.g. A is preferred to B is preferred to C). Utility theory can also describe choice between risky options, but ordinal rankings are not enough to do the job: the *cardinal* value of utility must be specified. The following example illustrates this point. Suppose that a wealthy relative offers you a choice between three sums of money, 0, X, and 100 dollars, and you know that X is between 0 and 100 dollars. Which sum would you choose? If you had a riskless choice, you would choose 100 dollars, regardless of X's value; in other words, ordinal utility is enough to determine your choice. Suppose, now, that you had the choice between two lotteries. Ticket A gives you 0, X, and 100 with equal probabilities, and ticket B give you 0 and 100 with equal probabilities. Which ticket would you choose if X equals 0.01? Which if X equals 99.99? You show a remarkable lack of avarice if your choice does not depend on the value of X. In risky choices the value of X matters!

THE EXPECTED-UTILITY HYPOTHESIS

We pointed out in Chapter 1 that a currency of optimization can be useful only if it allows us to rank all the decision-maker's options. How can a lottery ticket (or, more generally, a probability distribution), which only represents the probabilities of various outcomes, be ranked? Intuitively, it may seem that the mean (or expected) rewards from lottery tickets might be used to rank them. In the second example discussed in the previous paragraph, if $X = 80$, then ticket A would be ranked above ticket B [since $(0 + 80 + 100)/3 = 60$ is greater than $(0 + 100)/2 = 50$]. Unfortunately, human (and animal) decision-makers do not always pick the option with the highest mean reward. Does this not destroy the idea of using means to rank lottery tickets? Not according to utility theory; if you prefer ticket B to ticket A (even if A has a greater mean), it is because you do not assign 80 units of extra value (or utility) to 80 additional dollars. There is, a utility theorist would argue, a *utility function* $U(r)$ (see section 5.2) that relates your perceptions of value to monetary rewards (r), and your preferences indicate that the expected utility you would derive from lottery B is greater than the expected utility you would derive from lottery A:

$$\frac{U(0) + U(80) + U(100)}{3} < \frac{U(0) + U(100)}{2}$$

or

$$U(80) < \frac{U(0) + U(100)}{2}.$$

According to utility theory, the utility function describes your preferences. The expected-utility hypothesis (the basic premise of utility theory) is the claim that, regardless of how complicated your preferences are, there exists a utility function, the expectation of which ranks all probability distributions (or lottery tickets). The expected-utility hypothesis may not hold if your choices are inconsistent or irrational (e.g. if you prefer A to B and B to C, but you prefer C to A, then you are inconsistent, and no utility function will describe your behavior; see Raiffa 1968, DeGroot 1970).

In short, when we apply utility theory to foraging behavior we suppose that foragers make consistent and rational decisions, and so we can use utility functions to describe the preferences of foragers. In risky situations the fundamental currency of optimization becomes "maximize expected utility" (instead of "maximize utility," as it was in Chapter 5).

UTILITY, RISK PRONENESS, AND RISK AVERSION

A *risk-prone* forager prefers a probability distribution of rewards to the distribution's mean value with certainty. (The term risk-prone is confusing because it means that a forager prefers risk, not that a forager is susceptible to risk. However, we use the phrase because of its wide acceptance.) A *risk-averse* forager prefers the mean amount with certainty to the probability distribution. These definitions of risk proneness and risk aversion are restrictive, because they deal with a hypothetical choice between a certain pay-off and a gamble with the same mean value. If the decision-maker faces a more complicated choice, for example, if the alternatives vary in both mean and variance, then it may choose the more variable alternative even if the forager is technically risk-averse.

Different types of utility functions are associated with risk proneness and risk aversion. We illustrate with three examples the importance of the utility function's shape.

1. *A Straight Line.* In a straight line (or linear) utility function utility increases with reward at the same rate, irrespective of the amount of reward already obtained: every additional unit of reward adds the same amount of utility (Fig. 6.1[A]).

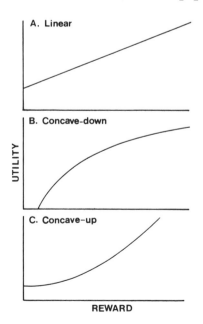

Figure 6.1 Three utility functions with different concavities. (A) A linear utility function. (B) A concave-down utility function. (C) A concave-up utility function.

Figure 6.2 An illustration of why concave-down and concave-up utility functions "act" differently under risky conditions. In both cases the forager is offered a fair lottery in which the same number of units of reward may be lost as may be won. Winning and losing have equal probability ($\frac{1}{2}$). In the concave-down case the forager might lose more "utility" than it might win, so a forager with a concave-down utility function would prefer not to gamble. In the concave-up case the forager might win more than it might lose, so it prefers to take the gamble.

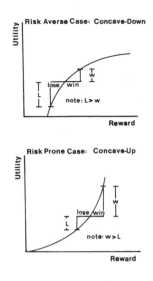

2. *A Concave Down Curve* In a concave-down (negatively accelerated) utility function each unit of reward is valued *less* than the last (Fig. 6.1[B]). A forager with such a utility function wants less as its own needs are increasingly satisfied. This is the "law of diminishing returns."
3. *A Concave-Up Curve.* In a concave-up (positively accelerated) utility function each additional unit of reward is *more* valuable than the last (Fig. 6.1[C]). This is the mirror image of the previous case.

Each of these three utility functions corresponds to a different type of risk sensitivity: a linear utility function predicts risk indifference, a concave-down utility function predicts risk aversion, and a concave-up utility function predicts risk-proneness. Figure 6.2 gives a graphical derivation of these claims in a simple case. When the utility function is linear, we can safely ignore risk sensitivity, because mean-reward maximizing is equivalent to expected-utility maximizing for linear utility functions. On the other hand, when the utility function is not linear, we must know the shape of the utility function to predict preference in risky situations.

How General Is the Law of Diminishing Returns?

In Chapter 5 we referred only to the concave-down utility function, and we argued, as have most economists, that additional rewards must *eventually* add less value. It is not, however, difficult to imagine a utility function with *increasing returns*. Suppose that an animal needs to travel 20 meters to get to a water hole. Stopping at 18 meters does as much to slake the animal's thirst as stopping at 19 meters (that is, neither reward helps at all), but the last meter from 19 to 20 changes things a lot. The claim that utility functions *always* show diminishing returns is unjustified; at least part of a utility function may be positively accelerated.

The law of diminishing returns, and its incorrect extension to the claim that utility is always negatively accelerated, has had a hypnotizing effect on some utility theorists (see Friedman and Savage 1948). Initially, economists rejected expected-utility theory's description of human choice behavior, because they *knew* that utility functions were always negatively accelerated, and that humans willingly gamble. They claimed that utility theory failed because it could only explain risk avoidance. In an important paper Friedman and Savage (1948) pointed out that the law of diminishing returns need not apply to the entire utility function. Most of the utility function may be concave-down, and risk aversion may be more common than risk proneness, but commonness is not the same as generality.

MORE AND LESS RISK SENSITIVITY

Suppose that a risk-averse forager has a utility function (defined over food reward) that is concave down at every point. Is the simple fact that this beast is risk-averse an adequate description of its behavior? The answer is generally no, because some foragers are more risk-averse than others. Consider the curves shown in Figure 6.3(A) and (B). Both curves go through the points $[0, U(0)]$ and $[20, U(20)]$, because $U_a(0) = U_b(0)$ and $U_a(20) = U_b(20)$, but the curve in panel A (or simply curve A) is straighter than the curve in panel B (curve B). Suppose that the owners of each utility function face a gamble in which 0 and 20 are equally likely. Because both utility functions are concave down, both individuals would rather have 10 units (the mean of the gamble) with certainty. However, these individuals do have different preferences: they have different certainty equivalents. The *certainty equivalent* is the smallest certain amount the forager would trade for the gamble (Fig. 6.3). It measures (in mean reward) the amount that a risk-averse individual would be willing to give up to have certainty. Curve A has a higher certainty equivalent (8) than curve B (4). The owner of curve B will give up more for certainty than the owner of the straighter curve A. In other words, the owner of curve A is less risk-averse than the owner of curve B. The difference between the certainty equivalent and the mean of the probability distribution is a measure of risk aversion called the *risk premium*. For example, in the simple 0–20 gamble $[P(0) = P(20) = \frac{1}{2}]$ above, if the forager was indifferent between the gamble and 4 units, then it would have a risk premium of 6 units (see Fig. 6.3[B]). However, the forager with the less bowed utility function (Fig. 6.3[A]) might have a risk premium of only 2 units. The risk premium represents the amount the risk-averse forager will pay for certainty. From the analyst's point of view, the risk premium has the disadvantage of depending on the gamble being considered.

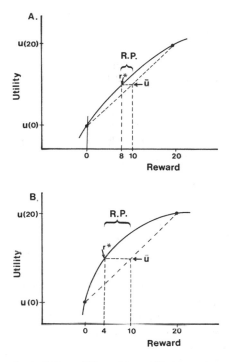

Figure 6.3 A comparison of more (B) and less (A) risk-averse utility functions. The expected utility of the gamble $[P(0) = P(20) = \frac{1}{2}]$ can be found by connecting the points $[0, U(0)]$ and $[20, U(20)]$ with a line. The expected utility is the point $(\bar{U})$ on this line corresponding to the mean reward (10 units). The certainty equivalent can be found by constructing a horizontal line of height $\bar{U}$. The reward that specifies the point (r^*) of intersection of this horizontal line with the utility function $[\bar{U} = U(r^*)]$ is the certainty equivalent. (A) The certainty equivalent is 8 reward units (the risk premium [RP] is 2 units) in the less risk-averse case. (B) The certainty equivalent is 4 reward units (the risk premium is 6 units) in the more risk-averse case.

The relative straightness of Figure 6.3's two curves suggests an alternative measure of risk aversion. The degree of departure from linearity (as measured by the second derivative) can be used as a measure of risk sensitivity. The *Arrow-Pratt measure of absolute risk aversion* is the second derivative of the utility function divided by the negative of the first derivative: $R_A = -U''(r)/U'(r)$. The Arrow-Pratt measure of risk aversion is a local measure; it tells us about the response to risk at a particular point along the utility function. There is an irony here, because by definition risk sensitivity treats problems in which knowledge of a particular point is not enough to predict preference (probability distributions spread likelihood over many possible amounts of reward).

What happens to the forager's risk sensitivity if we leave all features of the gamble the same but move it generally upward by shifting the mean reward? There are two important cases. First, if $R_A(r)$ decreases with r, a forager is said to practice *decreasing risk aversion*. A forager practicing decreasing risk aversion might pay a lot for a reduction in variance when it is just scraping by (low r), but it might be almost indifferent to variance when the reward on offer is large (high r). Second, if $R_A(r)$ is constant regardless of r, a forager is said to practice *constant risk aversion*. For example, a forager practicing constant risk aversion will pay the same sum to reduce the variance by one unit regardless of the mean reward's size.

RISK AND VARIANCE?

It is tempting to caricature the problem of risk as a trade-off between mean and variance; according to this view you might reread the previous section substituting "variance" for "risk." *Variance* has a special meaning; it is one of two parameters that characterize the normal distribution. All probability distributions have variance, but in most probability distributions, except the normal, mean and variance are closely related. The view that risk equals variance is intimately linked to the statistician's article of faith that given enough summing up (the central-limit theorem) the world is normally distributed. The principal models of risk taking in foraging theory assume underlying normal distributions. One problem with equating risk and variance is that the world may not be normal. Undaunted, we will deal mainly with variance in the remainder of this chapter. Economists have tried to arrive at more general definitions of riskiness. (See Hey 1979 for an elementary discussion of this and other risky matters. The general literature of measures of riskiness begins with a well-known paper by Rothschild and Stiglitz 1970).

THE BASIC MODELS AND RISK

In section 2.1 and Box 2.1 we pointed out that the prey and patch models assume long-term average-rate maximization. If total foraging time is fixed, then long-term average-rate maximization is equivalent to maximizing *mean* energy gain; in utility jargon, this means that the basic models assume risk indifference (linear utility functions) over energy gains.

6.3 Risk-Sensitive Feeding Behavior

Are foragers risk-indifferent, as the basic models assume? The answer is a convincing no: foraging preferences depend on random variation in food reward as well as on the mean food reward. Several workers have shown this (Table 6.1), but the work of Caraco and his colleagues (Caraco et al.

Table 6.1
Partial summary of empirical results on risk

Study	Species	Varied quantity	Results
Leventhal et al. 1959	Rats	No. of food pellets	Risk-prone, but tendency decreased as mean amount was increased
Pubols 1962	Rats	Delay before reward	Risk-prone, but tendency decreased as mean delay was decreased
Herrnstein 1964	Pigeons	Delay before reward	Risk-prone
Davison 1969	Pigeons	Delay before reward	Risk-prone
Caraco et al. 1980b	Yellow-eyed juncos	No. of millet seeds	Risk-averse when energy budgets were positive, risk-prone when energy budgets were negative
Caraco 1981	Dark-eyed juncos	No. of millet seeds	As above
Real 1981 and Real et al. 1982	Bumblebees and wasps	Nectar reward in artificial flowers	Risk-averse
Waddington et al. 1981	Bumblebees	Nectar reward in artificial flowers	Risk-averse
Caraco 1983	White-crowned sparrows	No. of millet seeds	As Caraco et al. 1980b
Barnard and Brown 1985	Common shrews	Mealworms	As Caraco et al. 1980b
Battalio et al. 1985	Rats	No. of food pellets	Unable to repeat Caraco's result, but used highly skewed reward distributions
Wunderle and O'Brien 1986	Bananaquits	Nectar reward in artificial flowers	Risk-averse

1980b) on yellow-eyed juncos (*Junco phaenotus*) is a justly celebrated example. Caraco et al. presented two trays of seeds, each with an opaque paper cover, to a junco. The experimental bird chose a tray by hopping down to it and knocking off the paper cover with its beak. Once a choice had been made, the observer withdrew the tray that was not chosen to make sure that choices were mutually exclusive. One tray was risky, providing for example, 0 or 10 millet seeds with equal likelihoods, but the alternative tray *always* provided a fixed number of seeds. The number of seeds provided on this "certain tray" equaled the mean number of seeds provided on the "risky tray," (5 seeds). A junco that consistently chooses the risky

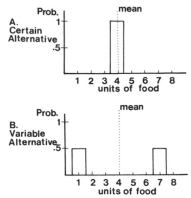

STANDARD RISK SENSITIVITY EXPERIMENT

Figure 6.4 This illustration shows the standard design of risk sensitivity experiments. The forager is offered a choice between (A) a certain alternative and (B) a probabilistically determined alternative with the same mean.

tray is risk-prone, and a junco that chooses the certain tray is risk-averse. This is the standard form of risk sensitivity experiments (Fig. 6.4).

Caraco et al. (1980b) measured the utility functions of their juncos, and the measured utility functions told them something about their juncos' risk sensitivity (Fig. 6.1). Measuring a utility function is a straightforward but time-consuming procedure. Suppose that an experiment aims to measure the utility function in the range 0 to 10 seeds, and assume that more seeds are better. Since utility functions can be transformed linearly without affecting their description of choice behavior, the end points can be arbitrarily assigned values, for example, $U(0) = 0$ and $U(10) = 1$. By offering many certain choices paired with the risky choice $[P(0) = P(10) = \frac{1}{2}]$, the experimenter can "titrate" to find the certainty equivalent of this risky choice. Suppose these experiments show indifference between 4 seeds with certainty and the risky choice; by definition, then, the expected utility of the risky choice [which is known: $\frac{1}{2} = U(0)/2 + U(10)/2$] must be equal to the utility of 4 seeds. This gives the first measured point on the utility function $(4, \frac{1}{2})$. This process can now be repeated to find the number of seeds that have a utility of, for example, 0.25, by determining the certainty equivalent of the lottery $P(4) = P(0) = \frac{1}{2}$. Figure 6.5 shows two utility functions that Caraco et al. measured for the same individual junco.

Caraco et al.'s most important finding was that juncos with positive energy budgets (i.e. those given enough food to allow them to meet their daily energy requirements) were risk-averse (they had concave-down utility functions), while juncos with negative energy budgets were risk-prone. The expected-energy-budget rule summarizes this result: If your expected (daily) energy budget is positive, then be risk-averse; if your expected (daily) energy budget is negative, then be risk-prone. (We put the word *daily* in

Figure 6.5 Estimated utility functions of an individual yellow-eyed junco (*Junco phaenotus*). Reward is the number of millet seeds offered per choice. (A) The estimated utility function when the junco's expected energy budget was positive. The negative acceleration of this function indicates risk aversion. (B) The estimated utility function when the junco's expected energy budget was negative. The positive acceleration of this function indicates risk proneness.

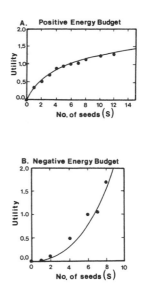

parentheses to indicate that Caraco and most workers following him have considered daily—as opposed to, say, hourly or weekly—energy budgets.) This remarkable result has now been repeated using two other species of small granivorous birds (Caraco 1981, 1983), starlings (Lima, personal communication), and shrews (Barnard and Brown 1985). Some authors have not measured utility functions, but they have all shown the switch from risk aversion to risk proneness associated with a change in energy budget. However, one study (Battalio et al. 1985) using rats and highly skewed reward distributions was unable to repeat this effect.

6.4 Shortfall Models of Risk Taking: The Z-Score Model

One way to model Caraco's energy budget result, and other aspects of risk taking, is to minimize the probability of an energetic shortfall (Caraco 1980, Stephens 1981, Houston and McNamara 1982, McNamara and Houston 1982, Pulliam and Millikan 1982, Rubenstein 1982, Stephens and Charnov 1982, Caraco and Lima in press). The argument is simple: if a forager must meet a certain fixed requirement for survival, then the forager's response to risk affects its chances of meeting its requirement. For example, suppose that a forager must choose between two alternatives; one choice yields 7 calories with certainty and the other (a risky choice) yields 3 calories half the time and 9 calories half the time. A mean-maximizer would take the certain alternative, because it has the higher mean. If the forager needed at least 8 calories to survive, which alternative should it choose? The risky choice gives a $\frac{1}{2}$ probability of survival, but the certain choice is certain death!

Stephens (1981) and Stephens and Charnov (1982) have formulated the so-called z-score model. This model maximizes a small bird's probability of surviving the night. The bird gets its food in small bits throughout the day. The pay-offs from each bit are randomly and independently distributed, so the sum of these pay-offs (in calories) will be normally distributed. The z-score model claims that the bird's energy supply at dusk (S_0) is normally distributed, and that the bird has some behavioral control over the mean and variance of this distribution. This control might be exercised by choosing where to feed.

The mean of the daily reward distribution is μ and the variance is σ^2. The remaining component of the model is a fixed daily requirement R. The model seeks to maximize

$$P(S_0 > R) = P(\text{surviving the night}).$$

Since S_0 is normally distributed with mean μ and variance σ^2, the likelihood of survival can be found by converting R to a standard normal deviate, or z-score $[Z = (R - \mu)/\sigma]$.

$$P(S_0 > R) = 1 - \Phi(Z),$$

where Φ is the cumulative distribution of the normal. Most elementary books on statistics tabulate the value of $\Phi(Z)$. Fortunately, the details of Φ are not crucial. In fact, Φ always increases with Z: large Z's mean small probabilities of survival. Minimizing Z maximizes the probability of survival.

Now imagine a special case in which the bird must choose the shortfall minimizing variance at a fixed mean. What σ would a Z minimizer prefer? The effects of σ on Z can be summarized as follows:

$$\frac{dZ}{d\sigma} = \frac{-(R - \mu)}{\sigma^2} = \frac{\mu - R}{\sigma^2}; \tag{6.1a}$$

$$\frac{dZ}{d\sigma} > 0 \quad \text{if } \mu - R > 0 \text{ or } \mu > R, \tag{6.1b}$$

$$\frac{dZ}{d\sigma} < 0 \quad \text{if } \mu - R < 0 \text{ or } \mu < R. \tag{6.1c}$$

A Z minimizer's risk sensitivity (risk equals variance) depends only on the relative sizes of its expected supply (μ) and its requirement (R). If expected supply exceeds the requirement, variance *should be reduced* (because increasing σ increases Z [expression 6.1b]). If expected supply is less than the requirement, variance *should be increased* (because increasing σ decreases Z [expression 6.1c]). This result is consistent with Caraco's expected-energy-budget rule.

The z-score result is more specific than the expected-energy-budget rule, because it predicts that the smallest variance possible should be chosen when μ exceeds R, and that the largest variance possible should be chosen when R exceeds μ. We call this the extreme-variance rule to distinguish it from the expected-energy-budget rule. This result is important because it shows that both risk proneness and risk aversion follow from a simple hypothesis of shortfall avoidance. We do not need to concoct a different explanation for each (as some economists, e.g. Hirshleifer 1966, have proposed).

This extreme-variance rule is a property of normal distributions. It will only be as general as normal distributions. There are important reasons that lead us to think that normal distributions of food supply might be common in nature. First, the central-limit theorem suggests that total supply will be normal when it is the sum of many independently distributed "acquisition events." Second, Stephens and Charnov (1982) have presented more rigorous arguments to show that the net gains from "a standard foraging process" (as envisioned in Chapter 2) will often be normal. However, even if normality is common, it is not general. Shortfall-minimizing tactics can be calculated for non-normal distributions (see McNamara and Houston 1982, Caraco and Chasin 1984).

COMPLICATED FEASIBLE SETS

Up to now we have considered only cases in which the forager chooses from a range of variances at a fixed mean. Many of the behavioral "choices" available to a forager will change both the mean and the variance. The z-score model can be extended to treat this problem. A particular value of Z, for example Z^*, defines many combinations of mean and standard deviation (μ and σ) that give the same probability of a shortfall. Borrowing terminology from Chapter 5, the line

$$\mu = R - Z^*\sigma, \quad \text{for } Z^* \text{ fixed} \tag{6.2}$$

defines an indifference curve over μ-σ combinations. Equation 6.2 is the slope-intercept form of a line on a μ-σ plot (Fig. 6.6). The requirement is the μ intercept and the slope is $-Z^*$. Since small Z^* means large $-Z^*$, a Z minimizer (survival maximizer) would prefer to have a (σ, μ) pair lying on a line radiating from R with a higher slope.

We can represent the forager's feasible choices as a set in (σ, μ) space. The "best" (minimum likelihood of shortfall) (σ, μ) pair in this set can be found by constructing *the* line of highest slope that intercepts the μ axis at

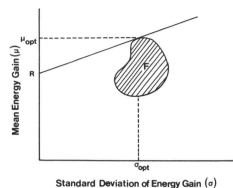

Figure 6.6 The (σ, μ) pair that minimizes the probability of an energetic shortfall is found by constructing the line of highest slope originating from R (the requirement) on the μ axis and passing through the feasible set (F).

R and passes through the feasible set. The point(s) of intersection of this line with the feasible set is/are the shortfall-minimizing (σ, μ) pair(s). Figure 6.6 illustrates this graphical solution. The extreme-variance rule (discussed above) fits neatly into this framework (Fig. 6.7).

RISK SENSITIVITY AND PATCH RESIDENCE TIME

Stephens and Charnov (1982) have worked out a simple example of how the z-score model might be applied to patch-leaving decisions. The first step is to calculate how the patch residence time affects the mean and standard deviation. Figure 6.8 presents the feasible set (of means and standard deviations) that a forager can choose by deciding how long to stay in a patch. The patch residence behavior of a shortfall-avoiding forager can be worked out using this feasible set. For example, a forager with a large requirement should leave the patch before a mean maximizer would, and a forager with a small requirement should stay longer. Because of the unusual shape of the feasible set, the behavior of a shortfall avoider will often be similar to the behavior of a mean maximizer.

The z-score model serves as a good starting point for a discussion of shortfall-avoiding models. In the remainder of this section we deal with objections and extensions to this model.

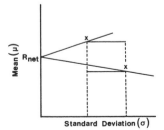

Figure 6.7 The graphical solution of the extreme-variance rule. The figure shows two horizontal lines that represent ranges of standard deviation from which the forager can choose at a constant mean. The higher line is above R, and minimum standard deviation gives the line of highest slope. The line below R shows that the maximum standard deviation gives the highest slope.

Figure 6.8 Patch residence time and the risk of starvation. (A) The curve shown is the feasible set of mean-standard deviation pairs that a forager can choose by choosing a patch residence time. The t's are the patch residence times (t_i's marked on the graph show how increasing residence times change the mean and standard deviation, $t_{i+1} > t_i$), and t^* is the patch residence time at which the mean energy gain is maximized. (B) This graph shows the optimal patch residence time for a short-fall avoider. If the forager's requirement is large, it should not stay as long as a mean maximizer would, but if the forager's requirement is small, it should stay longer. Notice that because of the ellipse-like shape of the feasible set, shortfall-avoiding patch residence times will usually be near the mean-maximizing patch residence time.

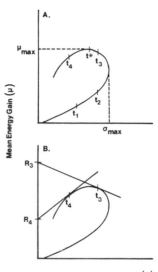

THE LAZY-L

The bird's energy supply can be compared to the position of a randomly moving particle. At intervals the bird's energy supply moves up or down depending on the random results of its foraging. The particle's path can be represented as a point (n, S), where n is the number of intervals left in the day and S is the energy supply at time n. In the z-score model any path is acceptable if the supply exceeds the requirement at dusk. However, the particle's path may drop below some critical level before dusk: the bird might die in the afternoon. Compared with a particle's "random walk," death in the afternoon is like encountering an absorbing boundary: death is a boundary that cannot be recrossed. Instead of viewing starvation as a hurdle that must be surmounted at the end of the day, this model views it as an absorbing boundary shaped like a reclining L (Fig. 6.9).

Stephens (1982) has studied this problem and found two interesting results (McNamara and Houston 1982 analyze some different aspects of

Figure 6.9 The lazy-L model of death in the afternoon. Time left in the day is on the abscissa, and energy supply is on the ordinate. The hatched bars indicate states that are equivalent to death. Death is a kind of absorbing boundary.

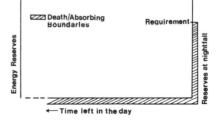

this problem). First, both the amount of time left in the day and the present energy supply determine whether the presence of the lower absorbing boundary affects the z-score model. As might be expected, when the present supply is large the lower absorbing boundary is less important. The lazy-L is most important when dusk is far off and becomes less important as dusk (i.e. the time by which the forager must have R) approaches. This effect is analogous to paying bills. Keeping a healthy balance in your checking account (avoiding the lower absorbing boundary) is more important than paying a bill when it first arrives, but as the due date gets closer the impetus for paying the bill increases. The mathematical reason for this effect is that the lazy-L can only affect the forager's behavior if there are paths that lead both to overnight survival and to death in the afternoon. There are few such paths just before dusk, but many when dusk is still a long time off.

Second, the lazy-L has an obvious but important effect on the extreme-variance rule. Suppose that the forager cannot expect to survive the night. The extreme variance rule predicts that it should gamble and choose high variance. Should it do this even if it is very close to death in the afternoon? The answer is no. Here, neither the highest nor the lowest possible variance is usually the best. The lazy-L suggests that animals might prefer an intermediate level of risk as a result of the competing goals of immediate survival and eventual need. This is an interesting result because it parallels the theory of investments (i.e. portfolio theory; see Coombs 1969, Caraco 1982), which suggests that human investors should prefer an intermediate level of risk.

ENERGETIC CARRY-OVER

In the z-score model survival is survival. It does not matter whether the forager survives the night by a hair's breadth or by a generous margin. Suppose that an insect larva requires a fixed amount of energy and nutrients to survive pupation. A surplus of energy or nutrients may be "carried-over" by the insect larva into adulthood. Energy carried over may be beneficial, because it allows earlier emergence, or greater gamete production.

The z-score model's zero carry-over assumption is equivalent to assuming that the forager's utility function is a simple step function. However, if energy carried over benefits the forager, then the utility function might increase sharply at the survival requirement but continue to increase smoothly after that. A carry-over function is the increasing part of the utility function for amounts of energy greater than the requirement for survival. A simple linear utility function shows the qualitative effects of carry-over.

In the "no carry-over" z-score model a forager whose expected reward equals its requirement should be indifferent to variance: the probability of survival is one-half regardless of variance, because the normal distribution is symmetric about its mean. When there is positive carry-over, a forager whose expected reward equals its requirement will always *prefer high variance*. It is still true that variance does not affect the likelihood of survival, but increasing variance increases the expected utility achieved, given that the forager survives. This is because increasing variance gives greater weight to larger energy values; increasing variance also gives greater weight to low energy values, but this does not matter because all ways of falling below the requirement are equivalent. We call this tendency for the truncated mean to be increased by variance, and its consequences for forager risk taking, the "achieved mean effect" (following Stephens 1982).

Suppose that the forager expects to be just above its requirements. Figure 6.10 shows how variance affects expected utility in the special case of a linear carry-over function. The results are intuitively appealing, given the achieved mean effect. When the forager is offered a range of small variances (between 0 and σ^* in Fig. 6.10), it prefers the smallest one. When the forager is offered only large variances (all its choices are larger than σ^*), it prefers the largest possible variance. When the variances offered are neither all less than σ^* or all greater than σ^* either the largest or smallest variance can be the best choice, depending on exactly what range is offered. When the variances offered are all small, the increase

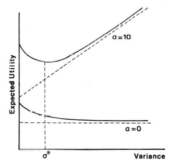

Figure 6.10 The achieved mean effect is illustrated. These are plots of expected utility as a function of standard deviation for the case in which the mean reward is greater than the requirement. The variable α is the slope of the linear carry-over function. The $\alpha = 0$ case is the same as the basic z-score, and expected utility decreases with increasing standard deviation, as it should. However, in the $\alpha = 10$ case expected utility decreases at first (achieving a minimum at σ^*) but then begins to increase. The implications of this are discussed in the text.

in achieved mean that could be had by increasing variance is small compared with the increased likelihood of death that would result. On the other hand, when the variances offered are all so large that the forager suffers a significant likelihood of death in any case, the increases in achieved mean from increasing variance may outweigh the increases in the likelihood of death. The linear carry-over case makes the achieved mean effect more powerful than it really is. In most natural situations we suspect there would be *eventual* negative acceleration. This would reduce the significance of the achieved mean effect. McNamara and Houston (1982) have studied an instance of energetic carry-over in which the achieved mean effect is not important.

THE VALUE OF VACILLATION

The z-score model has a worrisome implication. It predicts that any forager pressed enough to take a risk-prone alternative has a low probability of survival (lower than $\frac{1}{2}$, since the z-score must be positive to predict risk proneness, and $1 - \Phi(z)$ is less than $\frac{1}{2}$ for all positive z's). This prediction arises because the z-score model allows the forager to make only one decision: being risk-prone in the morning requires being risk-prone in the afternoon. If we consider a sequence of decisions, then risk proneness is not so dangerous (Houston and McNamara 1982, McNamara 1983, 1984). We discuss "sequential risk-taking" in the next chapter.

6.5 A Descriptive Model of Risk Taking

Oster and Wilson (1978), Caraco (1980), and Real (1980a, 1980b) have presented a descriptive model of feeding preferences. They propose that a crude but robust model for combining mean and variance is to maximize

$$\mu - k\sigma^2, \tag{6.3}$$

where μ is mean, σ^2 is variance, and k is a constant measuring the undesirability of variance. Real (1980a) has called this the variance discounting model, because it maximizes the mean discounted by a certain amount for variance. Real, the main proponent of this model, offers no guidelines about what quantities μ and σ^2 are the mean and variance of: they could equally well be energy or egg number. This model is a biological translation of what economists call the Markowitz-Tobin model of risk aversion. Economists use this model mostly for its analytical convenience, rather than for its general explanatory powers. Box 6.1 outlines the logic of this model.

BOX 6.1 DERIVATION OF VARIANCE DISCOUNTING

There are two ways to derive the variance discounting model of risk sensitivity: an exact and an approximate derivation.

EXACT DERIVATION

Our exact derivation follows Caraco (1980). Suppose that the forager's utility function is always concave-down and shows risk aversion. The utility function $-e^{-tX}$ $(t > 0)$ satisfies these requirements (a negative utility may seem strange, but linear transforms of utility, like adding one, make no difference to the results). The variable X is a random variable representing food reward. The expected utility is $-E[e^{-tX} | f(x)]$, where $f(x)$ is the probability distribution function of X. Those familiar with theoretical statistics will recognize $E[e^{-tX} | f(x)]$ as the moment generating function of $f(x)$. This is extremely handy, because statisticians have tabulated the moment generating functions for many probability distributions. If $f(x)$ is a normal distribution, then the expected utility is

$$-e^{-t[\mu - (t\sigma^2)/2]},$$

which always increases with increasing

$$\mu - \tfrac{1}{2}t\sigma^2;$$

or, equivalently, we can maximize

$$\mu - k\sigma^2, \text{ where } k = t/2.$$

Thus variance discounting follows from a utility function with constant risk aversion and underlying normal distributions.

APPROXIMATE DERIVATION

Suppose, following Real (1980a), that we do not know the shape of the utility function, $U_1(X)$. We can still use a Taylor's series to approximate $U_1(X)$ by expanding around a known point. The most convenient x value to expand around is μ. But before going on, we transform the utility function by adding $\mu - U_1(\mu): U(X) = U_1(X) + \mu - U_1(\mu)$. This yields the handy property that $U(\mu) = \mu$. Taking a Taylor's series about μ gives

$$U(X) = \mu + U'(\mu)(X - \mu) + \frac{U''(\mu)}{2!}(X - \mu)^2 + \ldots;$$

now we take the expectation of both sides,

$$E[U(X)] = \mu + \frac{U''(\mu)}{2}\sigma^2 + \ldots.$$

BOX 6.1 (CONT.)

Since $E(X - \mu) = 0$ and $E[(X - \mu)^2] = \sigma^2$, if we drop all higher order terms, then this becomes variance discounting, where $k = U''(\mu)/2$, and k will be negative if U is concave-down.

Since a Taylor's series can be used to approximate any differentiable function, it is tempting to conclude, as Stephens and Charnov (1982) have, that this derivation lifts the restrictions of underlying normality and constant risk aversion, which are required in the exact derivation. This is a rash conclusion, because utility functions with changing curvature (e.g. the z-score) and asymmetric probability distributions can make this a poor approximation.

One problem with this approximate derivation is that the Taylor's series works best *in the neighborhood* of the mean. Since probability distributions spread likelihood over many points, by definition, many possible realizations of reward may not be in the mean's neighborhood. In some situations the greater the spread, or variance, the worse the Taylor's series approximation. Figure B6.1 shows this effect. A constantly risk-averse, $U(X) = -e^{-x}$, forager faces gamma distributed rewards. We plotted the difference between exact expected utility and the Taylor's series approximation of the utility function when the mean was constant and the variance was increased. As the variance increases, the variance discounting formulation becomes a worse approximation (Fig. B6.1). We must therefore use variance discounting carefully: the safe alternative is to stick to conditions in which the exact derivation is reasonable.

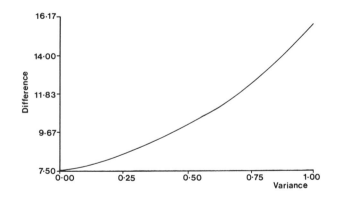

Figure B6.1   The figure shows the difference between the expected utility and the Taylor's series approximation of the utility function. The case shown is a utility function with constant risk aversion $[U(X) = -e^{-x}]$ and gamma distributed rewards with mean equal to 2. The difference between the exact expected utility and the Taylor's series approximation is plotted for increasing levels of variance. The approximation gets worse as variance increases.

Variance discounting versus the z-score. Variance discounting differs from the z-score model in two ways. First, variance discounting predicts constant risk aversion, while the z-score predicts decreasing risk aversion (see section 6.2 and Stephens and Paton 1986). Second, variance discounting is an *a posteriori* optimization model: risk-taking behavior must be observed to fit the constant of risk aversion. In contrast, the z-score model (like other shortfall-avoiding models) is an *a priori* optimization model, because the experimenter can estimate the forager's energy requirements, independent of observed risk taking. However, if there is exponential energetic carry-over (section 6.4), then the resulting model has the z-score and variance discounting as special cases (here, variance discounting's k measures the value of energy carried over). Thus the two models sometimes may be complementary.

6.6 Impulsiveness, Hunger, and Time Discounting

The major limitation of long-term rate maximizing is that it ignores the temporal pattern of energy intake: infinite energy gains tomorrow are irrelevant to an animal that will starve unless it finds another prey item before dusk. The z-score model begins to consider the importance of "acquisition pattern" by arguing that energy acquired before some critical "day of reckoning" is, sometimes, more important than longer-term gain. However, some evidence suggests that even shorter-term gains may be important. Many psychologists have studied whether animals prefer large, delayed rewards or small, immediate rewards (see Staddon 1983 for a review). These operant experiments suggest that animals sometimes sacrifice long-term rate maximizing for immediate gains (Box 6.2). Is this a response to risk?

Psychologists generally explain this "impulsiveness" without explicit reference to "risk." They argue that rewards now are fundamentally more valuable than rewards in the future, just as a human investor values 5 dollars today more than 5 dollars tomorrow. Kagel et al. (1986) argue that a delayed reward has a lower *present value* than a future reward, because something—a predator, a mate, an aggressive conspecific—might interrupt the forager during the delay: the longer the delay, the less likely that the reward really will be obtained at the end. In principle, this "discounting by interruptions" explanation only requires mean maximizing, because Kagel et al.'s "present value" is the same as the expected reward (Box 6.2).

However, Snyderman (1983b) has shown that hungrier animals are more impulsive (hungry animals are less selective in many situations; see

BOX 6.2 PRESENT VALUE AND CHOICE

Green et al. (1981) have studied the choice problem shown in Figure B6.2. Their pigeons were presented with a mutually exclusive choice every τ seconds. One choice led to a small reward (m_s seconds; they controlled reward size by controlling the seconds of access to food) after a delay of δ_s, followed by a post-feeding delay of Δ. The other choice led to a larger reward (m_L seconds) after a longer delay of δ_L. The time required to take a small ($\delta_s + m_s + \Delta$) was always equal to the time required to take a large ($\delta_L + m_L$). A rate-maximizing model predicts (see section 3.2) that the large choice should always be preferred. Green et al. showed that the delay to small (δ_s) affects choice (Fig. B6.2[B]):

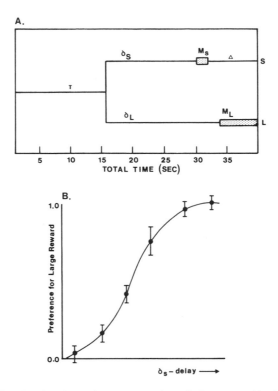

Figure B6.2 (A) A schematic representation of Green et al.'s (1981) "self-control" experiment: τ is the fixed time between choices, δ_s is the delay before a small reward, δ_L is the delay before a large reward, Δ is the post-feeding delay after a small reward, and m_s and m_L are the small and large rewards, respectively, measured as seconds of access to food. (B) Typical data from Green et al.'s experiment. As δ_s increases, the preference for the larger delayed reward increases.

BOX 6.2 (CONT.)

when small rewards occurred immediately after the "choice" point, they were preferred over large, delayed rewards. Green et al.'s pigeons flouted the first five chapters of this book!

Kagel et al. (1986) have tried to explain this "impulsive" behavior using standard economic arguments. Delayed reward has, they argue, a lower *present value* than immediate reward. Kagel et al. argue that the present value of a reward $[PV(m, \delta)]$ is (1) an increasing function of reward size (m) and (2) a decreasing function of delay (δ). When there is no delay, present value equals reward size $[PV(m, 0) = m]$.

If $\pi(\delta)$ is the instantaneous decrease in value at delay δ, then

$$PV(m, \delta) = me^{-\int_0^\delta \pi(z)\,dz}.$$

According to standard economic arguments $\pi(\delta)$ is a constant (π); thus

$$PV(m, \delta) = me^{-\pi\delta}.$$

However, if π were constant, only changes in the difference in delays ($\delta_L - \delta_S$) could change preference, but Kagel et al. point out that the evidence shows that absolute delay affects preference even when the difference is constant. Thus $\pi(\delta)$ must be a decreasing function of delay.

WHAT CAUSES TIME DISCOUNTING?

A human investor values 5 dollars today more than 5 dollars tomorrow for two reasons: (1) an interruption, some "act of God," may prevent the investor from collecting tomorrow's fiver and (2) 5 dollars today can be put to immediate use: it can earn a day's worth of interest before tomorrow's 5 dollars can be collected. Kagel et al. show that the interruptions model (in which rewards are weighted simply by the probability that they will still be there after delay) produces the decreasing $\pi(\delta)$ function that is required. However, Green et al.'s pigeons experienced no interruptions. Another potential explanation is that food gain "reduces" the cost of a deficit (see McFarland and Houston 1981). If cost per unit time depends on the forager's present food supply, then, by reducing the cost sooner, a small, immediate reward may reduce the cost of the deficit more than a larger, delayed reward would.

Schoener 1971), which hints at a risk-sensitive component to time discounting. If foragers collect fewer delayed rewards (as Kagel et al. suggest), then delays will affect both the mean reward per choice *and* the variance in rewards per choice: immediate rewards will be almost certain, and delayed rewards will be risky. In any model with decreasing risk aversion (section 6.2) we might expect that hungrier, but not starving, foragers will have stronger preferences for the low-variance immediate rewards, as Snyderman reports.

Time discounting might explain the apparent disagreement between the psychological and ecological results shown in Table 6.1. Although behavioral ecologists have offered a choice of variable and fixed *amounts* to their experimental subjects, psychologists have offered variable and fixed *delays*. According to the theory of time discounting, the utility of a reward decreases with delay in a concave-up fashion; hence, risk-proneness over delay is expected (Houston and McNamara, in preparation).

Although hunger has long been known to influence foraging behavior (see Schoener 1971), conventional foraging theory has not dealt directly with hunger. Some authors have argued that foragers use their hunger to estimate the "habitat rate of intake" (Charnov 1976a). However, in Snyderman's foraging experiments hungry pigeons were less selective (more impulsive), even though both hungry and less hungry birds had plenty of experience with the experimental encounter rates. The z-score model gives foraging theory a new view of hunger: it might be represented as increased requirements, or as decreased expected gain. The z-score formalizes the appealing idea that hunger represents a "threat of shortfall," and changes in the behavior of hungry foragers can be viewed as reactions to this threat. The z-score model itself probably cannot account for all the documented effects of hunger, but it points to the direction foraging theory might take in its effort to explain them.

6.7 Summary

A forager faces a problem of *risk* when it must choose between known probability distributions of reward, as opposed to choosing between certain, or riskless, options. Predicting preferences in risky situations requires cardinal measures of utility.

Foraging animals are not indifferent to risk, as mean-maximizing models require. Foragers show both risk aversion (preference for certainty) and risk proneness (preference for variability). Some foragers are risk-averse when they have positive energy budgets but risk-prone when they have negative energy budgets. The z-score model proposes that this switch in risk sensitivity occurs because it minimizes the probability of an energetic shortfall. The z-score model has some serious limitations. Many authors have tried to explain risk-sensitive preferences using a linear combination of mean and variance.

Shortfall-avoiding models show how the temporal pattern of energy gain might be important, and they show how hunger can be added to foraging theory.

7

Dynamic Optimization: The Logic of Multi-Stage Decision Making

7.1 Introduction

The prey model of Chapter 2 solved for the best value of p_i, the probability of attacking prey type i upon encounter. The model assumes that the optimal probability of attack (p_i) is constant: it does not change as a function of the forager's hunger or experience. Models in which the decision variable is constant are called "*static optimization*" *models*. In *dynamic models* a forager can change its decision from time to time: it might apply one probability of attack on the first encounter and another on the second. Dynamic optimization allows the decision variables to depend on such factors as experience, hunger, and body size, and these variables may in turn depend on previous decisions. Dynamic optimization allows for the possibility that a decision made today may affect tomorrow's decision: it finds optimal paths or trajectories, as opposed to the optimal points found by static models.

REVISITING THE Z-SCORE

The z-score model (presented in the previous chapter) specifies the levels of energy reserve at which it is optimal to choose high and low variance options (minimizes the probability of an energy shortfall). The z-score finds these energy reserve conditions using conventional static optimization, but energy reserves may change from time to time. They may even be changed by previous variance choices. Can the z-score be interpreted as giving a dynamic rule: At any time (t) choose the minimum variance ($\sigma_{\min}$) if the expected gains from now until dusk $[\mu(t)]$ exceed the expected requirements from now until dusk $[R(t)]$? The static techniques we used to find the z-score rule assumed that a constant variance was to be chosen: i.e. if the forager chose high variance in the morning, then it must choose high variance in the afternoon. Thus, strictly speaking, z-score cannot be extended in this way (Houston and McNamara 1982).

The interaction between risk taking and energy reserves should really be represented by a dynamic or multi-stage decision process. The forager's variance choices might be represented by the list $\sigma_1, \sigma_2, \sigma_3, \ldots, \sigma_{n-1},$ σ_n; σ_i is the standard deviation chosen at the ith interval. If the probability of a shortfall could be expressed as a function of the members of this list [e.g. $\rho(\sigma_1, \sigma_2, \sigma_3, \ldots, \sigma_{n-1}, \sigma_n)$], then in principle static techniques could be used to find the list of optimal σ_i's. This could be done by solving the n simultaneous equations $\partial \rho / \partial \sigma_i = 0$, but this conventional approach throws away a helpful piece of information. The decision represented by σ_n is taken *after* all the other σ_i's have been chosen. The list $\sigma_1, \sigma_2, \sigma_3, \ldots, \sigma_{n-1}, \sigma_n$ is more than a garden variety vector of decision variables; it is a sequence of decisions.

Without knowing how the first $n - 1$ σ_i's were chosen, suppose that the forager has reached the last decision with energy reserves S_n. What σ_n should be chosen? Since only one decision is now in question, static optimization is appropriate, and the conventional z-score results apply:

$$\text{choose } \sigma_n = \sigma_{\min}, \quad \text{if } R - S_n < \mu; \tag{7.1a}$$

$$\text{choose } \sigma_n = \sigma_{\max}, \quad \text{if } R - S_n > \mu, \tag{7.1b}$$

where R is the gross daily requirement and μ is the expected gain per interval. Furthermore, the conventional z-score specifies the probability of a shortfall as a function of S_n:

$$\rho_1^*(S_n) = \begin{cases} \Phi[(R - S_n - \mu)/\sigma_{\min}], & \text{if } R - S_n < \mu \\ \Phi[(R - S_n - \mu)/\sigma_{\max}], & \text{if } R - S_n > \mu \end{cases} \tag{7.2}$$

It is generally true that the solution from the static, one-decision case is the "last step" solution for dynamic optimization. The function $\rho_1^*(S_n)$ and the conditions (7.1) yield the optimal pay-off and optimal decision implications for all possible values of S_n (the energy reserves before the last decision is taken). This function can be used to find the optimal variance when there are two decisions left. The energy supply before the nth decision (S_n) is related to the energy supply before the $n - 1$th (S_{n-1}) by the equation $S_n = S_{n-1} + Y$, where Y indicates a random variable drawn from a normal distribution with mean μ and standard deviation σ_{n-1}. The probability of a shortfall can now be expressed as a function of S_{n-1} and σ_{n-1}:

$$\rho_2(S_{n-1}, \sigma_{n-1}) = \int_{-\infty}^{\infty} \rho_1^*(S_{n-1} + y)\phi(y \mid \mu, \sigma_{n-1}) \, dy, \tag{7.3}$$

where $\phi(y \mid \mu, \sigma_{n-1})$ is a normal probability density function with parameters μ and σ_{n-1}. The optimal value of σ_{n-1} and the minimum probability of a shortfall, given two decisions remaining, for any value of S_{n-1} can now be found by applying ordinary single-decision static techniques to the problem of minimizing expression (7.3). Unfortunately, expression (7.3) is difficult to minimize, but static optimization technique is not important here. It is more important to recognize the way that a dynamic optimization problem can be solved by working backwards from the last step and performing static optimization at each step. McNamara (1983, 1984) has shown, using a different argument, that the conventional z-score rule can be applied at the next-to-last step, giving the following:

$$\text{choose } \sigma_{n-1} = \sigma_{\min}, \quad \text{if } R - S_{n-1} < 2\mu; \tag{7.4a}$$

$$\text{choose } \sigma_{n-1} = \sigma_{\max}, \quad \text{if } R - S_{n-1} > 2\mu; \tag{7.4b}$$

and

$$\rho_2^*(S_{n-1}) = \begin{cases} \rho_2(S_{n-1}, \sigma_{\min}), & \text{if } R - S_{n-1} < 2\mu \\ \rho_2(S_{n-1}, \sigma_{\max}), & \text{if } R - S_{n-1} > 2\mu \end{cases}. \tag{7.5}$$

Although the extreme-variance rule is still optimal at the next-to-last step, the probability of a shortfall when two decisions remain is different from the probability of a shortfall when only one decision remains. Table 7.1 shows a comparison between $\rho_1^*(S_n)$ and $\rho_2^*(S_{n-1})$. The table shows that choosing $\sigma_{\max}$ is less dangerous when there are two decisions left to make. In the static problem if the forager chooses high variance at dawn, it must choose high variance until dusk, since only one decision is being considered. The dynamic model allows the forager to become conservative (switch to low variance) if its gambles pay off, and the flexibility increases the forager's probability of avoiding a shortfall.

Table 7.1

Probabilities of death when a risk-prone decision is optimal[a]

Difference[b] of requirement and expected gains	Probabilities of death	
	1 step left	2 steps left
0.00	0.500	0.338
0.25	0.550	0.377
0.50	0.599	0.416
0.75	0.650	0.456
1.00	0.691	0.496
2.00	0.841	0.648
2.75	0.915	0.744

[a] These are the results of numerically integrating equations (7.2) and (7.5), assuming a mean of one, a maximum standard deviation of two, and a minimum standard deviation of one.

[b] The requirement is always greater than the expectation when risk proneness is predicted. The difference is expressed as a proportion of the maximum standard deviation.

BOX 7.1 DYNAMIC PROGRAMMING

This box derives the fundamental equations of dynamic programming. The general problem is to find a sequence of decision variables $u_1, u_2, \ldots, u_n$, where each u_i is a decision made at time i. The state at time i is called x_i. Energy supply was the state variable in the sequential z-score. A "next-state function" (McFarland and Houston 1981) defines the changes in state as a function of the previous state and the previous decision. The "next-state function,"

$$x_{i+1} = G(x_i, u_i), \tag{B7.1a}$$

is said to represent the system's dynamics, because it describes how the state and the decision influence the next state. The currency of maximization is assumed to have the convenient form

$$C = \sum_{i=1}^{n} H(x_i, u_i). \tag{B.7.1b}$$

In this deterministic case (stochastic elements can be added to equation [B7.1a] in at least three ways; see Jacobs 1974) the initial state (x_1) and the list of decisions $(u_1, u_2, \ldots, u_n)$ specify the value of C, since all other x_i's can be found using the next-state function. This is shown by writing C as a function of only the initial state and the sequence of u_i's, $C(x_1, u_1, u_2, \ldots, u_n)$. Here, $f_n(x)$ is the best possible pay-off given the initial state x $(x_1 = x)$:

$$f_n(x) = \max_{u_1, u_2, \ldots, u_n} [C(x, u_1, u_2, \ldots, u_n)]. \tag{B7.1c}$$

The maximization indicated can be performed in arbitrary order, so we choose to maximize with respect to u_1 last:

$$f_n(x) = \max_{u_1} \left\{ \max_{u_2, \ldots, u_n} [C(x, u_1, u_2, \ldots, u_n)] \right\}. \tag{B7.1d}$$

The first term of C, referring back to equation (B7.1b), is independent of later decisions $(u_2, \ldots, u_n)$, so

$$\max_{u_2, \ldots, u_n} [C(x, u_1, u_2, \ldots, u_n)]$$

$$= H(x, u_1) + \max_{u_2, \ldots, u_n} \left(H[G(x, u_1), u_2] + H\{G[G(x, u_1), u_2], u_3\} + \ldots \right).$$

By analogy to $f_n(x)$, the maximized term in square brackets is called $f_{n-1}[G(x, u_1)]$. It has the same interpretation as $f_n(x)$, and it is the best possible pay-off from a sequence of $n - 1$ steps beginning with state $x_2 = G(x, u_1)$. Substitution shows that $f_{n-1}[G(x, u_1)]$ is related to $f_n(x)$ in the following way:

$$f_n(x) = \max_{u_1} \{ H(x, u_1) + f_{n-1}[G(x, u_1)] \}. \tag{B7.1e}$$

BOX 7.1 (CONT.)

This expression relates every pair of consecutive decisions, since it was developed for arbitrary n. It relates f_1 to f_2 as well as it relates f_{25} to f_{26}. The last-step solution,

$$f_1(x) = \max_{u_1}[H(x, u_1)], \tag{B7.1f}$$

and equation (B7.1e) are the only equations necessary (in principle) to solve a dynamic optimization problem. Equations (B7.1e) and (B7.1f) are sometimes called the fundamental equations of dynamic programming. Expression (B7.1e) shows how dynamic programming evaluates a decision in terms of its immediate effect $[H(x, u_1)]$ *and* its effect on the value of future decisions $\{f_{n-1}[G(x, u_1)]\}$.

A solution reached by dynamic programming consists of finding the functions $f_i(x)$ in steps of increasing i (i.e. working backwards in time) and using the function $f_i(x)$ to find the optimal decision variable u^*_{i+1}. In static optimization workers are often content to find the u^*_{i+1}. In dynamic programming the u^*_{i+1} cannot be found until $f_i(x)$ is known, because $f_i(x)$ specifies the future effects of any choice of u_{i+1}.

THE PRINCIPLE OF OPTIMALITY AND DYNAMIC PROGRAMMING

In principle expression (7.5) could now be used to solve the three-step case (find the optimal σ_{n-2}), but this two-step case is enough to illustrate the basic approach to multi-stage decision making. It illustrates all the general features of dynamic optimization problems and their solutions. The most important thing to notice is the way the problem is approached by moving backwards in time. We must know the optimal behavior at the last step to figure out the optimal behavior at the next-to-last step; because the optimal behavior at the last step tells us how the results of the next-to-last step should be evaluated. This backwards approach reflects Bellman's (1957) principle of optimality: "An optimal policy has the property that, whatever the initial state and initial decision are, the remaining decisions must constitute an optimal policy with regard to the state resulting from the first decision." The principle of optimality is the most fundamental and general idea in dynamic optimization. Our reanalysis of the z-score applied this principle.

Mathematicians call the "brute force" application of the principle of optimality, which our reanalysis of the z-score model illustrates, *dynamic programming*. Box 7.1 derives the so-called fundamental equations of dynamic programming. These equations show how, in a dynamic problem, a decision is evaluated in terms of its immediate effect *and* its effects on the value of future decisions, Dynamic programming is usually difficult, and it seldom leads to an analytical solution. There are simpler techniques for solving dynamic problems, but none is as general as dynamic programming. If all else fails, dynamic programming and a computer will usually produce results.

7.2 Solving for Decision Functions: The PMP

So far we have discussed the problem of finding an optimal sequence of discrete decisions, u_1^*, u_2^*, ..., u_n^*. A continuous dynamic optimization problem attempts to find a function of time $u^*(t)$ that describes the best decision path or trajectory. The optimal decision trajectory, $u^*(t)$, can be found by dynamic programming. The fundamental equations of dynamic programming, expressions (B7.1e) and (B7.1f), can be modified for the continuous case (Jacobs 1974 presents discrete and continuous cases side by side).

Dynamic programming is difficult in both discrete and continuous cases, but in the continuous case there is an important mathematical technique that *can* make dynamic solutions easier to find. This technique is called Pontryagin's Maximum Principle (PMP). PMP finds $u^*(t)$ by combining static optimization with the solutions of ordinary differential equations. Many texts prove PMP (Dixit 1976 presents a clear development), but it is necessary to at least state it here, because the results in the remainder of this chapter require a rudimentary knowledge of PMP.

PMP: PMP maximizes the *currency integral*

$$\int_0^T F[x(t), u(t), t]\, dt. \tag{7.6}$$

In this expression $x(t)$ is a state variable expressed as an unknown function of time, $u(t)$ is the decision function (for which a solution is sought), and t is time. Here x and u are scalars, but they could just as well be vectors. There will usually be a required condition $x(0) = b_0$, where b_0 is some constant. To find the optimal decision function $u^*(t)$, we must know how the state (x) changes with time as a function of the state itself (x), time (t), and the decision variable (u); that is, we require an equation of the form

$$\frac{dx}{dt} = g[x(t), u(t), t]. \tag{7.7}$$

This function is a continuous version of Box 7.1's next-state function. The function g represents the "dynamics" of the system, and it is sometimes called the plant equation. This is a comparison to industrial control processes, in which the dynamics give the relationships between changes in state and the design of the "plant." The equation

$$H[x(t), u(t), \beta(t), t] = F[x(t), u(t), t] + \beta(t)g[x(t), u(t), t] \qquad (7.8)$$

is called the Hamiltonian. In this expression $\beta(t)$ is only a function of time and it is an introduced and unknown function (which acts like a Lagrange multiplier). It may be called the co-state variable, the auxiliary variable, or the multiplier. PMP states that the $u(t)$ that maximizes the currency integral can be found by a static maximization of the Hamiltonian, plus the following two ordinary differential equations:

$$\frac{d\beta}{dt} = -\frac{\partial H^*}{\partial x} [x^*(t), \beta(t), t], \qquad (7.9)$$

$$\frac{dx^*}{dt} = \frac{\partial H^*}{\partial \beta} [x^*(t), \beta(t), t]. \qquad (7.10)$$

The asterisks (*) above indicate that the partial derivatives are evaluated at the optimum $u^*(t)$.

An additional condition for maximizing the currency integral, often left out of biological descriptions of PMP, is the *transversality condition*. The transversality condition states that the co-state function $\beta(T)$ equals zero at the time horizon T. This condition holds when T is fixed and finite and $x(T)$ is not fixed. If T is not fixed and finite, or if $x(T)$ is fixed, then more complicated statements can still be made about the value of $\beta(T)$ [see Takayama 1974; T will not be fixed in time minimization problems, and some problems will have a fixed target state $x(T)$].

The plan for finding $u^*(t)$ by PMP is outlined below:

(i)   Find the value $u^*(t)$ that maximizes the Hamiltonian. This can be done by calculus or other static means. The result will give $u^*(t)$ as a function of $x^*(t)$, $\beta(t)$, and t. Substituting $u^*(t)$ back into H gives $H^*[x^*(t), \beta(t), t]$, which is no longer a function of $u^*(t)$.

(ii) The result of the first step is used in equations (7.9) and (7.10). The two resulting ordinary differential equations can (in principle) be solved for $\beta(t)$ and $x^*(t)$ as explicit functions of time.

(iii) Since $u^*(t)$ was found as a function of $x^*(t)$ and $\beta(t)$ in the first step, the explicit functions of time found in the second step can be substituted to find $u^*(t)$ as an explicit function of time.

Formally, PMP finds $u^*(t)$ as a function of time alone (Box 7.2). The function $u^*(t)$ may not be biologically interesting, because it hides information about how the decision is related to the state. For example, a typical state

BOX 7.2 THE PMP: SOME SIMPLIFICATIONS AND AN EXAMPLE

TWO USEFUL RESULTS

A solution using PMP can sometimes be simplified, because it can be argued that the Hamiltonian is constant along the optimal path. The Hamiltonian is constant along the optimal path when time enters the Hamiltonian only through other variables, that is $H[x(t), u(t), \beta(t), t] = H[x(t), u(t), \beta(t)]$. The chain rule shows this result:

$$\frac{dH^*}{dt} = \underline{\frac{\partial H^*}{\partial x}\frac{dx}{dt} + \frac{\partial H^*}{\partial \beta}\frac{d\beta}{dt}} + \frac{\partial H^*}{\partial t}. \tag{B7.2a}$$

Substituting equations (7.9) and (7.10) shows that the underlined part of this equation is zero. If $\partial H/\partial t$ is also zero, then the Hamiltonian is constant along the optimal path.

The decision variable may be the rate of change of state variable $u = dx/dt = \dot{x}$. For example, foraging behavior may be thought of as controlling the rate of change of energy supply. Here the Hamiltonian is

$$F[x(t), \dot{x}(t), t] + \beta(t)\dot{x}(t)$$

(F is the integrand of the currency integral). Differentiating with respect to the decision variable,

$$\frac{\partial F}{\partial \dot{x}} = -\beta(t),$$

and

$$\frac{\partial H^*}{\partial x} = \frac{\partial F}{\partial x}.$$

These equations can be substituted into expression (7.9) to yield

$$\frac{d}{dt}\left\{\frac{\partial F}{\partial \dot{x}}[x^*(t), \dot{x}(t), t]\right\} = \frac{\partial F}{\partial x}[x^*(t), \dot{x}^*(t), t]. \tag{B7.2b}$$

This is called the Euler-Lagrange equation. The Euler-Lagrange equation was used to solve dynamic optimization problems before the PMP was known. Generally, equation (B7.2b) leads to a second order differential equation, and this can usually be solved. However, if any of the three arguments $[x^*(t), \dot{x}^*(t), t]$ does not affect F, then the solution is even simpler. If x does not affect F, then the right side of equation (B7.2b) is zero, and the equation can be integrated to give the solution

$$\frac{\partial F}{\partial \dot{x}}[x^*(t), \dot{x}^*(t), t] = \text{a constant of integration}. \tag{B7.2c}$$

BOX 7.2 (CONT.)

Dixit (1976) discusses the other two cases. Foraging theorists often study rates, so the Euler-Lagrange equation may be especially helpful.

AN EXAMPLE: THE SHORTEST PATH BETWEEN TWO POINTS

Biological applications of dynamic optimization seldom complete the solution by finding the control variable as a function of t (t is usually time). However, a completely worked example is the best way to show how a solution by PMP is supposed to work.

We have taken an example from Dixit (1976) that has no particular relevance to foraging, but as Dixit says "it has the great merit" that everyone knows the answer. Imagine a cartesian coordinate system with t (the equivalent of time in this example) on the abscissa and y (the state variable) on the ordinate. We want to find the function $y(t)$ that has the shortest path between $(0, 0)$ and $(1, 1)$. These points give initial condition $y(0) = 0$ and terminal condition $y(1) = 1$.

A result from elementary calculus says that the path length of any function $y(t)$ is

$$\int_0^1 \left[1 + \left(\frac{dy}{dt}\right)^2 \right]^{1/2} dt. \tag{B7.2d}$$

Thus, we want to maximize the negative of this expression. Now let the control variable be $u = dy/dt$. The Hamiltonian is

$$H = -(1 + u^2)^{1/2} + \beta u;$$

maximizing by setting $dH/du = 0$, we find that

$$u^* = \frac{\beta}{(1 - \beta^2)^{1/2}} \tag{B7.2e}$$

and substituting this back into the Hamiltonian, we find

$$H^* = -(1 - \beta^2)^{1/2}. \tag{B7.2f}$$

Now applying expression (7.9),

$$\frac{d\beta^*}{dt} = -\frac{\partial H^*}{\partial y} = 0.$$

Since β is a constant, u^* and H^* must also be constants (from equations [B7.2e] and [B7.2f]). We could have concluded that H^* was constant by observing that t does not enter H explicitly. Applying expression (7.10) shows that

$$\frac{dy^*}{dt} = \frac{\partial H^*}{\partial \beta} = \frac{\beta}{(1 - \beta^2)^{1/2}}.$$

BOX 7.2 (CONT.)

Integrating this expression and applying the initial and terminal conditions shows that

$$y^*(t) = t,$$
$$u^*(t) = 1,$$
$$\beta^*(t) = 1/\sqrt{2}.$$

This can be solved more easily using the Euler-Lagrange equation, since the control variable is the rate of change of the state variable. The integrand in expression (B7.2d) does not involve y, so we can use equation (B7.2c): $\dot{y} = dy/dt$, so

$$\frac{-\dot{y}}{(1 + \dot{y}^2)^{1/2}} = \text{a constant of integration} = c.$$

Solving this for $\dot{y}$ gives

$$\dot{y} = \frac{c}{(1 - c^2)^{1/2}} = \text{another constant, say, } k.$$

Solving this differential equation, we find

$$y^*(t) = kt + c_1,$$

where c_1 is another constant of integration. Applying the initial conditions, we find that

$$y^*(t) = t$$

as expected.

---

variable may be hunger, and a typical decision variable may be patch residence time. A biologist may be more interested in the relationship between hunger and patch residence time than in eliminating hunger from the relationship (by substituting in the relationship between hunger and time). Many biological PMP users have stopped at step (i), because they are interested in the relationship between decision and state variables. This may leave them with the awkward problem of finding a biological interpretation for the introduced function $\beta(t)$. [Like its cousin the Lagrange multiplier, $\beta(t)$ can be interpreted as the shadow price of a constraint; see Dixit 1976]. Although biological applications of PMP seldom complete the solution by finding $u^*(t)$ as a function of time, Box 7.2 presents a worked example, which does find $u^*(t)$, to help the reader understand the principle. Box 7.2 also discusses some situations in which PMP can be simplified.

7.3 Trade-offs and Dynamic Optimization

Dynamic foraging models usually focus on trade-offs (see Chapter 5 for a static approach to trade-offs). For example, the following dynamic trade-off problems have been studied: conflicts between feeding and predation (Milinski and Heller 1978, Heller and Milinski 1979, Gilliam 1982, in preparation); conflicts between feeding and territorial behavior (Ydenberg 1982 Ydenberg and Houston 1986); conflicts between feeding and drinking (Sibly and McFarland 1976); and even conflicts between production and reproduction in the social insects (Macevicz and Oster 1976).

Dynamic optimization is used to study trade-offs for two reasons. First, it seems natural to formulate trade-off decisions as functions of internal (or state) variables. The trade-off between feeding and drinking is conveniently formulated in terms of the internal deficits hunger and thirst. The modeler would like to know which combinations of hunger and thirst lead to feeding behavior and which combination lead to drinking behavior.

The second reason for studying trade-offs using dynamic optimization is that the time course of conflicting behaviors is often the phenomenon of interest. For example, ornithologists have observed that male great tits do much of their territorial defense in the morning, but they are also likely to be hungriest in the morning (Kacelnik and Krebs 1983). A model built to study the trade-off between feeding and territorial defense might try to predict the time course of feeding and defense.

GREAT TITS: TERRITORIAL DEFENSE AND FORAGING

In the spring male great tits, living in deciduous forests in Southern England, defend their territories by singing and patrolling in the tree tops. However, they do much of their feeding on the ground, so there is a conflict: when the male feeds it cannot pay full attention to territorial defense and advertising, but when it defends its territory it cannot readily feed.

Ydenberg (1982) and Ydenberg and Houston (1986) studied this conflict by training captive male great tits (individually housed in large aviaries) to collect food rewards from a feeder. This feeder delivered rewards when the bird hopped on a nearby perch; the pattern of food delivery was set as $B^{1/2}$ rewards obtained, on average, after B hops. This function mimics patch depression (section 2.3). At any point the bird could fly to a "reset" perch at the opposite end of the aviary. Landing on this perch reset the reward schedule in the feeding site (B was reset to 0). In this set-up an energy-maximizing great tit should forage in the patch and fly to the perch according to the pattern predicted by the marginal-value theorem (Chapter 2). However, Ydenberg and Houston (1986) placed an intruder (another male great tit in a cage) near the reset perch. The experimental male could see the intruder when it traveled to the reset perch, but it could not see the intruder when it foraged in the patch. This condition simulates the conflict that male great tits experience in the field: they can monitor and chase intruders when they travel between feeding sites, but not when they are on the ground in a feeding site.

Ydenberg and Houston argue that hunger might affect how a great tit balances feeding against defense: a well-fed great tit will be freer to allocate time to territorial defense. Moreover, this factor makes it a dynamic problem. Should a great tit feed now and defend later, or defend now and feed later? Suppose that $x(t)$ is a state variable that represents the great tit's hunger (or food deficit) at time t. In Ydenberg and Houston's experiment the food deficit is reduced at rate

$$\frac{dx}{dt} = -\frac{B^{1/2}}{B + \tau},$$ (7.11)

where B is the patch residence time, τ is travel time, and $B^{1/2}$ is the experimentally determined gain function (time is scaled so that one hop takes one time unit). A conventional way to handle this problem would be to argue that external factors, such as the number of intruders per hour, set the value of τ. If this were true, then changes in τ could be viewed as a simple constraint, and any given value of τ would set the optimal patch residence time according to the marginal-value theorem. Here, the rate-maximizing B equals τ, and the optimal rate of intake is $R_{max}(\tau) = (\tau^{-1/2})/2$.

Ydenberg and Houston argue, however, that both B and τ are decision variables. The costs of territorial defense increase with time spent feeding (B) and decrease with time spent traveling between patches (τ). Thus they argue that total costs per unit time are

$$k_B B + \frac{k_t}{\tau} + C(x),$$

where k_B and k_t are constants, and $C(x)$ is a function that specifies the cost per time unit of a food deficit of size x. Costs increase with hunger $[C'(x) > 0]$, but $C(x)$ does not depend on B or τ. Ydenberg and Houston seek to minimize total cost, so they maximize the currency integral

$$\int_0^T -\left[k_B B + \frac{k_t}{\tau} + C(x) \right] dt.$$ (7.12)

This minimizes total costs from time 0 to final time T, and it leads to the Hamiltonian

$$\beta(t)\left(\frac{B^{1/2}}{B + \tau} \right) - k_B B - \frac{k_t}{\tau} - C(x).$$ (7.13)

Calculus can be used to find the first order conditions for maximizing the Hamiltonian (i.e. set $\partial H/\partial B = 0$ and $\partial H/\partial \tau = 0$). This calculation shows that optimal B^* and τ^* values are related by

$$B^* = \frac{\tau^*}{k(\tau^*)^2 + 1},\qquad(7.14)$$

where $k = k_B/k_t$. In principle this result could now be used to find τ^* as a function of k_B, k_t, and $\beta(t)$, and equations (7.9) and (7.10) could be solved to find $\beta(t)$ as a function of time. Equation (7.14) could then be used to find B^* as a function of time. This solution is difficult to find, and Ydenberg and Houston are more interested in the relationship (7.14) between the two decision variables B^* and τ^* (Fig. 7.1[A]). In contrast to the marginal-value theorem, this model does not predict that patch residence time (B) always increases with travel time (τ); instead, this model predicts that patch residence time will reach a maximum at an intermediate value of travel time. The rate of energy intake in this model is always less than the rate the marginal-value theorem would predict if τ were regarded as a simple constraint. This reduction reflects the compromise between maximizing the rate of intake and controlling the costs of territorial defense.

The variable k ($= k_B/k_t$) measures the degree of "intrusion risk." If k_B increases or k_t decreases, the cost of "not defending" increases. Thus k should go up when more intruders threaten the territory. A higher k predicts a flatter relationship between B^* and τ^* (Fig. 7.1[A]).

This model does not determine how a great tit should allocate its efforts to feeding and defense during the interval 0 to T. We have not answered whether a great tit should delay feeding to shore up its territorial defense or delay defense to feed, because we have not found τ^* as a function of time. However, Ydenberg and Houston make the intuitively appealing argument that travel times should always increase as hunger decreases.

Ydenberg and Houston tested their predictions using the experimental set-up discussed above, and they found that travel times did increase as a great tit reduced its food deficit. They also found that patch residence time (B) reached a peak at an intermediate value of travel time (τ), and that the relationship between B and τ was generally lower when they experimentally increased intrusion risk (Fig. 7.1[B]).

SUNFISH: HABITAT CHOICE AND PREDATION

Gilliam (1982, in preparation; see also Werner and Gilliam 1984) has presented one of the more thorough and thoughtful analyses of a dynamic foraging problem. Gilliam analyzed how bluegill sunfish of different sizes should choose feeding habitats. A bluegill living in the open water is more likely to be preyed upon than a bluegill living in the weeds along the shore.

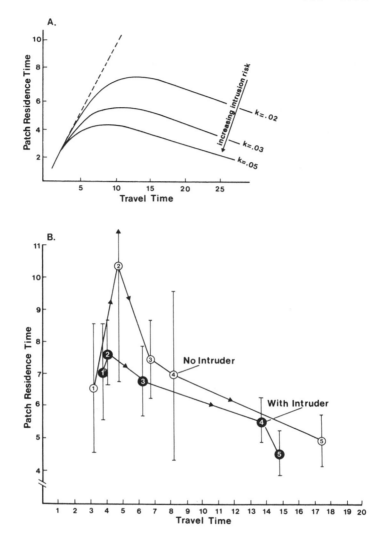

Figure 7.1 (A) The solid curves show the relationship between patch residence time and travel time predicted by Ydenberg and Houston's model. Increasing the threat to territorial intrusion (increasing the variable k) flattens the curve. The dashed, straight line shows the relationship predicted by the marginal-value theorem, which simply adds territorial defense time to travel time. At first Ydenberg and Houston's optima closely follow the rate-maximizing marginal-value prediction, but as travel times get larger, rate maximizing is sacrificed for territorial defense. (B) The results of Ydenberg and Houston's experiment. The open circles show a treatment without an intruder, and the black circles show a treatment with an intruder. Travel times increased as time in experimental conditions elapsed. Thus Ydenberg and Houston divided experimental time into quintiles. The circles marked 1 represent the mean patch residence time and travel time from the first quintile, and so on. The vertical bars are standard errors.

The open water is always more dangerous than the weeds, but large fish are less vulnerable in both weeds and open water. Figure 7.2 shows a hypothetical relationship between body size and mortality rate (m) in both weeds and open water. Staying in the weeds would always be a good plan, except that a bluegill can sometimes gain weight more quickly by feeding on plankton in the open water. Gilliam represents the quality of food in the two habitats by showing the hypothetical growth rate (g) that a fish of a given size achieves by feeding in each of the two habitats (see Fig. 7.2).

If the real mortality versus-size and growth-versus-size functions were like those in Figure 7.2, then it is obvious that staying in the weeds would be a good idea for any fish smaller than $\hat{s}$. But since the fish cannot stay smaller than $\hat{s}$ forever, what should it do once it becomes bigger than $\hat{s}$? Near $\hat{s}$ the growth rates are about the same. Why not stay in the weeds and experience the smaller mortality rate? If the fish does this, then it will be smaller next year and therefore will suffer more mortality then, because bigger fish are always less vulnerable to predators. In short, picking a growth rate today may determine tomorrow's mortality rate. The best solution can be found only by considering a lifetime trajectory of growth rates.

Gilliam supposes that the bluegill chooses the proportion of time spent in each habitat. If on a given day the bluegill chooses to spend $100p\%$ of its time in the weeds and $100(1 - p)\%$ in the open water, then this choice,

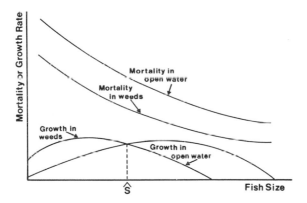

Figure 7.2 Hypothetical relationships for Gilliam's study of bluegill habitat choice. The figure shows relationships between growth rate and body size and between mortality rate and body size in two habitats (weeds and open water). The mortality rate decreases with body size in both habitats, and it is always lower in the weeds. The growth rate at first increases with body size, but it declines after reaching a maximum. At small body sizes the weeds both are safer and provide a higher growth rate. After body size $\hat{s}$ there is a conflict between the safest habitat and the habitat that provides the highest growth rate.

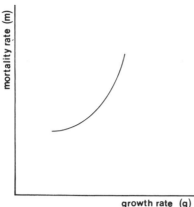

Figure 7.3 The feasible combinations of growth rate and mortality rate that a bluegill sunfish might choose at a given body size.

together with the functions shown in Figure 7.2, determines the mortality and growth rates it achieves on that day. From a mathematician's point of view, one can think of the bluegill's choosing a growth rate at each instant, because at a fixed size a growth rate uniquely specifies p, which in turn specifies a mortality rate. Figure 7.3 shows a hypothetical relationship between growth rate and mortality rate. Since growth rate and size determine mortality, the mortality rate can be expressed as a function of growth rate (g, the decision variable) and size (s), or $m(g, s)$.

Gilliam argues that (for a stable density dependent population) the optimal growth rate trajectory should maximize the following currency integral:

$$R_0 = \int_0^T \ell(t)b(t)\,dt.$$

In this equation, R_0 is the net reproductive rate, $\ell(t)$ is the survivorship to age t, and $b(t)$ is the fecundity at age t. The reader may recognize this as a standard expression from population biology. Here, T is a very large finite time horizon, and T must be chosen so that surviving beyond age T is impossible, $\ell(T) = 0$. Let $s(t)$ be the size at age t. The growth rate also will be a function of age. Thus the mortality rate can be written as a function of age alone, $m(t) = m[s(t), g(t)]$. The survivorship to age t will be

$$\ell(t) = e^{-\int_0^t m(y)\,dy} = e^{-D(t)}. \tag{7.15}$$

The two state variables are $s(t)$ and $D(t)$, and the corresponding dynamic equations are

$$\frac{dD}{dt} = \frac{d[\int_0^t m(y)\,dy]}{dt} = m(t) = m[s(t), g(t)], \tag{7.16a}$$

$$\frac{ds}{dt} = g(t). \tag{7.16b}$$

This leads to the Hamiltonian

$$e^{-D(t)}b[s(t)] + \beta_D(t)m[s(t), g(t)] + \beta_s(t)g(t). \tag{7.17}$$

The maximum with respect to the decision variable (g) can be found by differentiation

$$\frac{\partial H}{\partial g} = \beta_D(t)\frac{\partial m}{\partial g} + \beta_s(t) = 0 \tag{7.18a}$$

or

$$\frac{\partial m}{\partial g} = -\beta_s(t)/\beta_D(t). \tag{7.18b}$$

Since the Hamiltonian is not an explicit function of time, it must be constant along the optimal growth rate trajectory (see Box 7.2). Moreover, since the Hamiltonian has the same value at any point in time between 0 and T, its value can be calculated at any convenient age. The upper bound T is a convenient age, because here the co-state variables are both equal to zero [$\beta_s(T) = \beta_D(T) = 0$, by the transversality condition mentioned in section 7.2]. The upper bound was chosen so that $\ell(T) = 0$. Thus H must be equal to zero everywhere along the optimal growth rate trajectory. Setting the Hamiltonian (7.17) equal to zero and substituting the first order condition (7.18) leads to

$$\frac{\partial m}{\partial g} = \frac{m(s, g) + e^{-D}b(s)/\beta_D}{g}. \tag{7.19}$$

(Note that in this equation and the next one we no longer show variables as explicit functions of time t.) The choice of g must satisfy this expression at every point along the optimal growth rate trajectory. The juvenile case gives an especially elegant result. Since juveniles do not reproduce (by definition), $b[s(t)] = 0$. Juveniles should choose growth rates so that

$$\frac{\partial m(s, g)}{\partial g} = \frac{m(s, g)}{g}. \tag{7.20}$$

This leads to a graphical interpretation similar to the familiar graphs of the marginal-value theorem. Figure 7.4 shows this graphical interpretation. The comparison with the marginal-value theorem suggests another interpretation. The growth rate at size s that satisfies equation (7.20) *minimizes* the ratio $m(s, g)/g$. The choice of habitats under this minimization rule

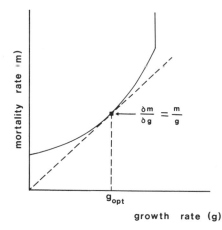

Figure 7.4 At a given body size the bluegill should pick that "growth rate-mortality rate" combination that minimizes the ratio of mortality rate to growth rate. A tangent drawn from the origin to the feasible "growth rate-mortality rate" curve intersects the curve at the point that minimizes this ratio.

can be a simple problem. If a juvenile bluegill must pick only one habitat when the "open water" growth rate is twice the "weeds" growth rate, then the open water should be preferred unless the "open water" mortality rate is more than twice as great as the "weeds" mortality rate. The model predicts that juveniles should trade off growth rates in proportion to mortality rates. In symbols the condition "choose habitat 1 if $m_1/g_1 < m_2/g_2$" is equivalent to the condition "choose habitat 1 if $g_2/g_1 < m_2/m_1$." For our purposes the "juvenile result" is a good stopping point. Gilliam (1982) discusses the conditions for mixed-habitat use by juveniles, the optimal growth rate trajectories for adults, and how his results are changed if the population is expanding or contracting.

7.4 Conclusions

Dynamic optimization is a powerful and general tool. We expect that it will become increasingly common in foraging theory. Unfortunately, few biologists understand dynamic optimization well, and this diminishes its value as a medium of communication. Moreover, many dynamic optimization problems are difficult to solve, and the approach to such problems often has been to force the biology into a soluble framework, instead of fitting the mathematics to the biology. Biologists have frequently used quadratic currency integrals and linear dynamics, because these (LQP) problems can be readily solved (see McFarland and Houston 1981, Jacobs 1974). If stochastic elements are added, dynamic optimization becomes even more difficult, leading to a body of stochastic foraging theory that relies almost solely on static optimization, and to a body of dynamic foraging theory that supposes a deterministic world. An important thrust of future work may be to bring these elements together (see Mangel and Clark 1986).

Despite this reservation, there are important cases for which nothing else will do. Dynamic models have made important points about trade-offs and their mediation by state variables (e.g. hunger, thirst, body size). Ydenberg and Houston's model shows how assuming that (between-patch) travel time is used for territorial defense changes the basic patch model. Gilliam's model puts foraging choices in a life history framework, and he finds a surprisingly simple result. McNamara's dynamic z-score model explains why more foragers survive risk taking than is predicted by the static z-score. McFarland and Houston (1981) review other applications of dynamic optimization to behavior.

7.5 Summary

Dynamic optimization is the study of multi-stage or sequential decision making; in the continuous case this means the study of decision paths or trajectories. A dynamic approach is called for when the "decision now" affects the "best decision" later. This situation usually occurs because the "decision now" affects the forager's future state, which in turn affects the economics of choice. The most general principle of dynamic optimization is Bellman's (1957) principle of optimality: "An optimal policy has the property that, whatever the initial state and initial decision are, the remaining decisions must constitute an optimal policy with regard to the state resulting from the first decision."

Dynamic optimization problems can sometimes be solved by the repeated application of the principle of optimality, beginning with the last decision and working backwards in time. This method is called dynamic programming. Dynamic programming is difficult and often requires a numerical solution.

In the continuous case a method called Pontryagin's Maximum Principle (PMP) can sometimes be used. PMP may lead to elegant solutions where dynamic programming only leads to a muddle, but not often. Dynamic foraging optimization has focused on trade-off problems with satisfying results. We present a case in which the marginal-value theorem fails because territorial defense conflicts with hunting within patches. We present a second model which considers the life history implications of foraging decisions. This model finds the optimal growth rate trajectory in cases in which achieving a high growth rate conflicts with predator avoidance, because the richest habitats may also be the most dangerous. The case of non-breeding individuals gives an elegant result.

8 More on Constraints: Rules of Thumb and Satisficing

8.1 Introduction

The word *constraint* is rapidly following in the footsteps of *strategy, fitness, drive,* and other catchwords whose popularity and vagueness spell their own demise. In the 1970s "the foraging ecology of the wren" became "foraging strategies of the wren," and in the 1980s the same research might be labeled "constraints on foraging in the wren." In Chapter 1 (section 1.5) we argued that constraints are all of those things that relate the decision variable to the currency. Constraints, of one form or another, have always been an important part of foraging theory. However, foraging theorists have ignored, for generality's sake, many constraints: how do bees perceive flowers, how do pigeons "choose" between alternatives? Foraging theory has eschewed these constraints as species-specific and mechanistic. The resurgence of interest in constraints is a reaction to this view. Constraint advocates argue that foraging theory's "generality" may evaporate into vagueness, unless it begins to deal with "species-specific and mechanistic" constraints. In this chapter we illustrate with some examples the main themes of constraint research.

In addition to the formal mathematical meaning of constraint (section 1.5), students of behavior use the term in at least four ways. In this book we distinguish between the following types of constraints: phylogenetic, developmental, behavioral, and time-budget constraints. Dawkins (1982) and Mayr (1982, 1983) discuss these and other uses of constraint in some detail. In Chapter 1 (section 1.5) we discussed constraints in broad terms, and we emphasized behavioral and time-budget constraints; these constraints limit the forager's behavioral repertoire. Now we turn to the two kinds of constraints which, rather than acting on the animal's present repertoire, limit the potential of the repertoire to change.

Phylogenetic constraints. Statements such as "given the pentaradial symmetry of echinoderms, it would be difficult for them to evolve into active pursuit predators" or "given their compound eye, insect predators could not evolve high visual acuity" refer to a class of phylogenetic constraints that Gould and Lewontin (1979) call "Bauplan constraints." As we point out in section 10.2, behavioral models of foraging deal with a finer level of analysis than that involving Bauplan constraints.

The same applies to a second kind of historical constraint, one arising from the principle of "least resistance": natural selection proceeds in small steps along the path of least resistance from existing material, and it does not plan ahead. This leads to some extraordinary contraptions: animals that breathe through their anus, nerves that take circuitous routes from the periphery to the central nervous system. Dawkins (1982) mentions the recurrent laryngeal nerve of a giraffe, which goes from the larynx to the brain via the base of the neck—hardly the solution that would be reached by a designer starting from scratch! Gould (1980) makes a similar point with his example of the panda's thumb. Although these examples provide excellent records of evolutionary history, they may appear to undermine the notion of optimal design. If the products of selection are jury-rigged contraptions, why talk of *a priori* models of good design? There are two related replies to this. First, optimality is not perfection. The "path of least resistance" argument that explains these imperfect contraptions is itself an optimality principle. Optimality principles can lead to contraptions. Second, the models we discuss do not ask why the panda's thumb is an enlargement of the radial seismoid; rather, they ask whether, given the thumb it has, the panda is an efficient forager, with the model explicitly defining "efficient." Of course the danger of reducing the level of analysis like this is that we might be studying only trivial minutiae.

Developmental constraints. The process of development can be viewed as a filter through which the range of possible phenotypes specified by the genome must pass (Alberch 1982, Maderson 1982). Oster and Alberch (1982) present a detailed theoretical analysis of this issue. They show how morphogenesis may be predisposed to move along certain routes simply because of the local forces arising from the physicochemical properties of cytoplasm. Their analysis shows that bifurcation is a general developmental principle: development translates a continuum of possible phenotypes into an "either/or" outcome. The cytoplasmic properties influencing development are themselves subject to selection, so Oster and Alberch's bifurcation surface cannot be viewed as an immutable constraint. As Maderson (1982) suggests, the only ultimate constraints may be the physicochemical properties of organic molecules.

8.2 Behavioral Constraints: Rules of Thumb

Shrews apparently do not have the sensory equipment to discriminate between prey of varying profitability (e/h). Instead, they discriminate by prey size, a cue that is usually, but not always, correlated with profitability (Barnard and Brown 1981).

Does this mean that shrews are suboptimal? Since all animals must solve their foraging problems using some mechanism (or *rule of thumb*), this question is generally important. There are two contrasting views. On the one hand, Janetos and Cole (1981) and Myers (1983) think that animals are equipped with "less than perfect" rules of thumb such as "take the largest," which, although efficient, are not as good as the optima described by simple foraging models. They argue for more work on suboptimal rules of thumb instead of on simple optimality models. On the other hand, Krebs et al. (1983) and Cheverton et al. (1985) see rules of thumb as refinements of the classical foraging models, into which more realistic (but probably more parochial) constraint assumptions are incorporated. In this view the animal using the rule "take the largest" may be optimizing within a larger number of constraints (e.g. sensory limitations for shrews) than the one using the rule "take the most profitable" (see also Dennett 1983). Cheverton et al. (1985) extend this argument and suggest that eventually design models based on optimality considerations and mechanistic models based on physiological bases of behavior could become unified as constrained optimization models. Although this research program has not proceeded far, there is a growing body of literature that tries to find "good but simple" rules of thumb.

8.3 The Performance of Rules of Thumb

Even if a forager's behavioral and physiological equipment limits its capacity to assess its environment, other rules of thumb may be compatible with the animal's equipment. This raises the question of whether some rules might do better than others.

The most extensive theoretical analyses of performance of rules of thumb deals with patch-leaving decisions, and as these rules have also been investigated experimentally, we take them as a case study (see also section 4.4). Iwasa et al. (1981), McNair (1982), and Green (1984) have analyzed slightly different aspects of the performance of patch-leaving rules. They all consider one or more of the following possible rules (Cowie and Krebs 1979): (1) a number rule "leave after catching n prey" (Gibb

1958); (2) a time rule "leave after t seconds" (Krebs 1973a); (3) a giving-up time rule "leave after g seconds of unsuccessful search" (Krebs et al. 1974); and (4) a rate rule "leave when instantaneous intake rate drops to a critical value r" (see section 4.4). It turns out that different rules do best in different conditions; a discussion of the three main conditions follows.

A single patch type. McNair (1982) compares the performance of number, time, and giving-up time rules for animals foraging in an environment characterized by a single patch type and stochastic encounters with prey within the patch. He shows that the nature of the gain function (Fig. 2.1) has an important influence on which rule does best. If the function rises steeply and then sharply flattens off, a number rule does best, since both of the other rules may involve staying in the patch during the less profitable part of the gain function. If, on the other hand, there is gradual resource depression, a giving-up time rule does best.

More than one patch type. McNair (1982) also considers the case in which there are two patch types that the forager can recognize before it enters them. If one patch type's gain function is always above the other, then the forager should have a longer giving-up time in the better type patch. In other words, if the forager recognizes patch quality before entering the patch, then it should be more persistent in a good patch, in the face of a run of bad luck, then it should in a poor patch.

Patch sampling. Green (1984) and Iwasa et al. (1981) analyze the more complicated problem in which patches vary in quality and cannot be recognized beforehand. Iwasa et al. studied the performance of three rules (number, time, and giving-up time) when the forager is faced with different distributions of the number of prey per patch. Intuitively, the best rule depends on the distribution of patch qualities (prior distribution of patch sub-types, Box 4.1). If all patches contain exactly two prey, for example, the best rule is "take two and leave." However, if half the patches contain four prey and the other half contain none, then a forager using the "take two and leave" rule will soon die of starvation. Here, a giving-up time rule is probably the best of the three. Generalizing from this simple example, Iwasa et al. show that (1) when each patch contains the same number of prey (but encounters with prey are stochastic), a number rule does best; (2) when the number of prey per patch has a high variance, a giving-up time rules does best; and (3) when the number of prey per patch follows a Poisson distribution, a time rule does best. These differences arise because the information value of a prey capture depends on the distribution of patch qualities. With a fixed number of prey per patch, each capture tells the forager that the patch is getting worse; with a high-variance distribution, a capture tells the forager that it may have hit the jackpot and it should stick around; and with a Poisson distribution, a capture gives no information about relative patch quality.

Green's approach is slightly different: he looks for the best possible patch-leaving rule and compares its performance with two of Iwasa et. al.'s rules—the fixed time and giving-up time rules. His model is not directly comparable to Iwasa et al.'s, since he assumes no patch depression. Green's best rule has the form "leave if less than p prey are found after n looks." Green assumes that prey occur in discrete slots, so hunting is just a series of "looks" in slots. Green's rule does better than a simple giving-up time rule, because it reduces the chance of leaving too early as a result of a run of bad luck (see also section 8.4), and because it allows a more accurate assessment of differences in patch quality. (It is also worth adding, parenthetically, that the giving-up time rule gets 90% of the best rule's pay-off in the cases Green examined.)

Figure 8.1 summarizes some of these points graphically. A giving-up time rule does best when there is a single patch type, when patch subtypes can be easily recognized, or when the distribution of prey per patch is clumped. Figure 8.1 shows a general representation of patch-leaving rules as suggested by McNamara (1982; see section 4.4). A variable v, the potential, declines from its starting value during unsuccessful search within a patch. When the forager encounters a prey, the variable jumps a certain amount. The size of the jump depends on the preceding duration of unsuccessful search, rising to a maximum as the time between captures increases. Thus the incremental effects are like resetting a timer to its starting value, with the amount of reset depending on how long the timer has been running. McNair's analysis of two discriminable patch types can be thought of as showing that the size of the increment is related to prey size, so that the predator stays longer in those patches containing larger prey (Figure 8.1[B]). When Green's rate-maximizing assessment rule is expressed in the same form, the key difference between it and a giving-up time rule is that the jump due to finding, say, the third prey item is the same, regardless of how much unsuccessful search has preceded the encounter (Fig. 8.1[C]). In other words, the increments are additive, so that a rapid burst of captures early in a patch visit will have a cumulative effect in increasing residence time. The two remaining rules can be represented in similar graphical terms: for the fixed time rule, the potential decays with time, independent of encounters, and for the fixed number rule, it decreases a certain amount with each capture (Figs. 8.1[D] and [E], Iwasa et al. 1981).

Increment-decay models such as the giving-up time and assessment rules have a long tradition as ethological models of behavior. In addition

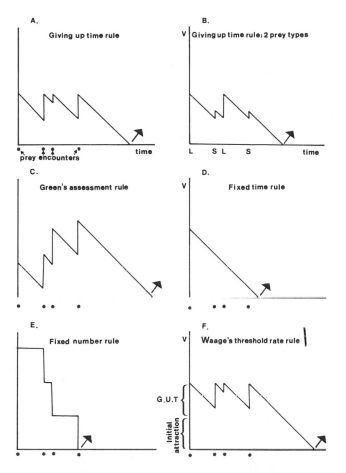

Figure 8.1   Graphical summary of various patch-leaving rules. (A) A simple giving-up time rule. The variable *v* declines with unsuccessful search, and it is reset to maximum value at each encounter. (B) The same as (A), but the size of the increment due to prey capture depends on prey size, with a different maximum for small and large prey. (C) Green's assessment rule. The increments are additive so that the rule responds to average, and not local, rate. (D) A fixed time rule in which encounters have no effect on the decay of *v*. (E) A fixed number rule in which the variable *v* declines a fixed amount after each encounter. (F) Waage's threshold rate model.

to their appeal as efficient rules of thumb (McNamara 1982) they bear an obvious resemblance to well-established neural mechanisms such as habituation and excitation and, as we shall see in the next section, they describe patch leaving in real animals.

8.4 Rules of Thumb: Experimental Evidence

Waage (1979) observed the parasitoid wasp *Nemeritis canescens* hunting for patches of *Plodia interpuntella* in a substrate of middlings. Waage found that the wasp followed an increment-decay model like Green's assessment model. Waage showed that two factors influence *Nemeritis's* patch residence time: (1) chemical stimuli from the hosts, which causes the wasp to probe with its ovipositor and engage in area-restricted search, and (2) successful oviposition, which causes the wasp to stay longer; it may increase the wasp's sensitivity to the host attractant chemical. Waage hypothesized that an interaction between decay in responsiveness and increments in responsiveness caused by oviposition determine patch residence time. These increments were not additive (as Green's rule predicts), but the giving-up time model (Fig. 8.1[A]) described the wasp's behavior best. (Waage allowed the wasps to encounter either 5 hosts in a burst at the beginning of a patch visit or 5 at 3-minute intervals. The wasps in the "burst" treatment left earlier than the wasps in the spread out treatment, so encounters do not have the cumulative effect that Green expects.) However, Waage's wasps did not use a simple giving-up time rule, because the last-capture-to-leave interval was often shorter than some of the earlier inter-capture intervals in the same patch visit. The additional factor proposed by Waage to "smooth out" stochastic variation in encounter rates early in the patch visit is the effect of chemicals produced by hosts: these set the variable v to a level above threshold, and only after this initial effect has worn off does the model behave like a simple giving-up time rule (Fig. 8.1[F]). This modified giving-up time rule produces patch residence times more like those expected from Green's rate model than like those predicted by an ordinary giving-up time rule.

It seems possible that increment-decay models, which are essentially elaborations of the giving-up time idea, can generally describe patch-leaving decisions. Roitberg and Prokopy (1984) and Ydenberg (1984) have used an approach similar to Waage's.

8.5 Rules for Switching on Concurrent Schedules

HILL-CLIMBING

Animals in Skinner boxes choose between two keys or levers offering different pay-offs (schedules of reinforcement), and often these choices are analogous to foraging decisions (Chapter 5, Staddon 1980, 1983, Kamil and Yoerg 1982, Lea 1982, Kamil 1983). In this section we discuss one aspect of the psychological literature on choice between alternative schedules of reinforcement: the moment-to-moment decision rule used by the animals. Staddon (1983) presents an extensive discussion of this problem, as well as other parallels between animal psychology and foraging theory.

Psychologists frequently study the choice between two so-called variable interval schedules (concurrent VI-VI, in the argot of the Skinner box aficionado). These VI schedules "set up" rewards at random time intervals, and rewards usually remain "set up" until the animal pecks the correct key. Obviously, the longer it has been since the last response to a key, the more likely it is that a reward will be received when the key is pecked. The achieved reward rate is independent of the response rate if the animal responds at any rate greater than one response per interval. Although it was not chosen because it bears a resemblance to natural foraging problems, the VI behaves like rapidly depleting food sources that are replenished in the forager's absence, for example, floral nectar (Kamil 1978) or insects washed up on a stream bank (Davies and Houston 1981). If rewards are set up with probability p during each time interval and the last response occurred n intervals ago, the probability of obtaining food, given a response, is

$$P(\text{food} \mid \text{response}) = 1 - (1 - p)^n,$$

which, if time intervals are small, can be approximated by

$$P(\text{food} \mid \text{response}) - 1 - e^{-\beta t},$$

where β is the rate at which rewards are "set up" and t is the time since the last response (Staddon 1983).

Hinson and Staddon (1983) present evidence that the choice rule for a pigeon in a concurrent VI-VI experiment is "pick the alternative with the higher probability of pay-off," a rule they call *hill-climbing* or momentary maximizing. The rule can be formalized as follows: Choose alternative A if $P(\text{food} \mid t_A) > P(\text{food} \mid t_B)$, where t_A and t_B are times since the last response on A and B respectively. If the two reinforcement schedules have identical set-up rates, then the rule reduces to "choose A if $t_A > t_B$." In general, the rule is: Choose A if

$$t_A > t_B(\beta_A / \beta_B). \tag{8.1}$$

This rule was tested by plotting the responses of pigeons on a graph whose axes are t_A and t_B (Fig. 8.2). Expression (8.1) gives the switching line in this graph: above the line a momentary maximizer should choose A, below the line it should choose B. Note that the rule and the data

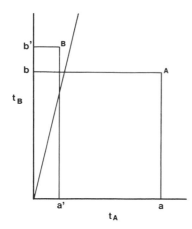

Figure 8.2 A graphical representation of hill-climbing in a concurrent VI–VI experiment. Here, t_A and t_B are the times since the last responses on A and B, respectively. The diagonal line divides the space into a region where the probability of getting rewards for a response is higher on A (below the line) and a region where it is higher on B (above the line). The point (a, b) should lead to choice A, and the point (a', b') should lead to choice B. As the schedule of A increases relative to B, the line rotates counterclockwise.

refer to steady-state behavior. Chapter 4 deals with the problem of acquisition, or how animals might find out about schedules of reinforcement.

Herrnstein and Vaughan (1980) and Vaughan and Herrnstein (1986) also have suggested that the rule for choice on concurrent schedules involves a form of hill-climbing, one that they call *melioration*. Melioration is less precisely defined than momentary maximizing, but it is similar. The verbal definition is as follows: "If the value of a (per unit time) exceeds that of b, relatively more time will come to be distributed to a" (Vaughan and Herrnstein 1986). Note that this definition does not specify the interval over which "value" is measured (Hinson and Staddon's hill-climbing refers to instantaneous probabilities), and that the form of response allocation is also not stated exactly ("more" as opposed to "all" in momentary maximizing). Despite these differences, melioration—like momentary maximizing— emphasizes *local rates of reinforcement* as the determinants of choice. Several studies have shown that local rates are important (Mazur 1981, Vaughan 1981, Vaughan et al. 1982). The animal must work on key A to set up rewards at key B, an arrangement which means that the highest overall reward rate could be obtained by working mainly at A and just hopping across to B to pick up the rewards. But since rewards are mainly obtained at B, an animal whose choice is guided by local as opposed to global rates of reinforcement will work mainly at B, and this is what psychologists find. However, this experiment is somewhat artificial (the only natural analogue that comes to mind is a fox tapping its paw at one entrance to a burrow and rushing round to collect the rabbit as it tries to escape from the other end—a problem that pigeons probably do not face in nature). While the contrived nature of the experiment may help us find the forager's choice mechanisms, just as bringing a bird into breeding season in midwinter with artificially long days reveals something about how breeding is controlled, it may limit the ability of the experiment to reveal why the rule is used. This brings us to the topic of matching versus overall maximizing.

MATCHING AND MAXIMIZING AS CONSEQUENCES OF HILL-CLIMBING

The matching law (Herrnstein 1970) describes overall (or molar) choice behavior in a general way:

$$T_A/T_B = w(r_A/r_B),$$

where T_i is the time allocated to alternative i, r_i is the reward rate *obtained* from i, and w is a weighting factor ($= 1$ for perfect matching). Thus the matching law states that relative time allocation matches relative rewards obtained. Although there is no doubt that the matching law describes many data, it is not clear whether its importance goes beyond this. It may reflect a moment-to-moment choice rule or a design feature favored by natural selection. We discuss these two possibilities in turn. First, it is intuitively obvious that momentary maximizing leads to an overall outcome similar to matching. Figure 8.2 shows that as the switching line rotates clockwise, the relative value of B increases, and so does the relative time allocated to B. Staddon (1983) reports that simulations of momentary maximizing show that it generally leads to slight undermatching: if the ratio of rewards obtained is 3:1, the time allocation will be a little less than 3:1. Herrnstein and Vaughan (1980) also suggest that their version of hill-climbing, melioration, leads to overall matching.

Second, there is considerable controversy about the relationship between matching and overall maximizing (Houston and McNamara 1981, Heyman and Luce 1979, Houston 1983). Without going into details, it is fair to say that matching and maximizing are exactly or nearly equivalent in some situations (the animal that matches also maximizes overall pay-off), but they are not equivalent for other schedules. In these cases animals approximate matching rather than overall maximizing. What is the relevance of this result to foraging theory? Suppose that hill-climbing is the basic rule of thumb. According to foraging theory, animals use this rule because it usually does well. If "doing well" means overall maximizing of pay-off, then the rule is used because of this consequence and not because of matching. The two outcomes are frequently similar, so the rule usually produces both. However, experiments can be designed so that the rule produces matching and not maximizing; according to foraging theory, the experimenter has tricked the animal into doing the wrong thing.

8.6 Satisficing and Constraints

"Instead of shoring up the idea of optimal design with more and more constraints, should we not simply abandon the idea that animals are optimal, and accept the idea that they are efficient but not optimal?"

Some authors extend this point by drawing a parallel between the concept of *satisficing* in decision theory and the notion that animals "do well, but they do not optimize" (Myers 1983, Krebs and McCleery 1984). This parallel is probably misleading. Decision theory uses satisficing (Simon 1956) and optimization as *purely descriptive* tools (*a posteriori* optimization): rational decision making can be described as utility maximization or, as Simon showed, as a satisficing process in which the decision-maker is satisfied after meeting some minimal requirement. As we pointed out in Chapter 1, foraging theory relies on *a priori* optimality arguments. Therefore, to justify talking about satisficing foraging theory instead of optimal foraging theory we would have to consider whether an argument from natural selection can be made for satisficing, just as we have argued for optimization. Such an argument can be summarized graphically by plotting fitness (say, reproductive success) as a function of performance according to some design criterion (say, rate of food intake). The argument for optimization is that fitness increases continually as a function of performance, at least to the maximum attainable value. The equivalent case for satisficing would be that fitness varies with performance according to a step function, so that above the threshold (Fig. 8.3) there is no relation between, say, food intake and reproductive success (Krebs and McCleery 1984). The threshold condition seems most unlikely to occur in nature, especially if one thinks about the effect of stochasticity. Suppose that there is a fixed requirement of food for survival, but the exact amount is unpredictable; the further the animal gets above the threshold, the less likely it is to fall within the stochastic range of the step change in fitness (see Myers 1983, Krebs and McCleery 1984 for other views of satisfaction).

However, let us accept for a moment the argument that there *is* a case for predicting that animals will be satisficers. How does this view fare as a recipe for research? We believe that it soon leads to a dead end because it too readily fits the facts. Take Krebs et al.'s (1977) observation that great tits, while they select more profitable prey, do not do so in the exclusive way predicted by simple diet models. One interpretation of this behavior would be that great tits are not optimizers, they are just efficient. This answer is confirmed by data, so no further investigation is necessary. A proponent of optimization, on the other hand, might consider whether

Figure 8.3 Hypothetical relationships between fitness and performance. The relationship in (A) might favor maximizing performance, and the relationship in (B) might favor satisficing.

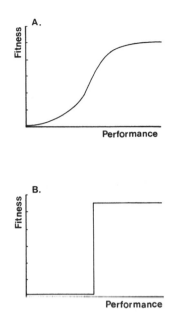

the original optimization model made incorrect assumptions about constraints. This line of argument would lead to a search for what the constraints might be. In the case under discussion, Rechten et al. (1983) showed that great tits make discrimination errors, and that this constraint is a component of the variation that leads to partial preferences (Stephens 1985, Box 2.3). Thus the "optimization with more constraints" approach has led to new experiments, and eventually to a deeper understanding of how prey selection works.

8.7 Concluding Remarks: Constraint versus Design

We conclude with two unresolved problems. The first, to which we have no simple solution, is that "one person's constraint may be another's design problem." The behavioral ecologist accepts the forager's sensory limitations as constraints, but neurophysiologists may well ask why the sensory system is designed the way it is. This difference reflects the hierarchical organization of living things. It is worth being aware of the problem, but it does not undermine the study of design at any particular level.

The second point seems to be a simple practical question. We have seen in previous chapters that there are two main ways to change foraging models: change the constraints or change the currency. How does the researcher know which to do? One obvious tactic is: if the interest is in evaluating alternative currency models, study systems with well-identified constraints; if the interest is in constraints, choose systems in which there is little doubt about the currency (or in which many currencies give the same predictions). With further thought, it becomes evident that constraint and currency are only partially separable issues: "constraint" is meaningless without something to constrain. The confounding of constraint and currency may reflect an important feature of nature: animals are a mess of competing goals and complex limitations. This chapter shows how currencies help us evaluate mechanisms: for example, we ask which of a set of simple patch-leaving rules does best. Optimality techniques may be powerful tools for studying nature precisely because they allow us to combine constraint and currency.

8.8 Summary

Foraging theory is beginning to pay more attention to the limitations on and mechanisms of foraging behavior. Some workers ask about rules of thumb: are there simple rules that do well despite their simplicity? Patch-leaving rules have been studied extensively in this context. Waage's work with ovipositing parasitoids is an elegant example of how deductions can be made about rules of thumb.

Some psychologists have taken a similar approach to explaining choice behavior. Hill-climbing and melioration are two rules of thumb that sometimes describe choice behavior. These rules may lead to so-called "matching."

Some workers argue that "satisficing" explains foraging behavior as well as "optimality" does. However, there is no reason to suppose that animals satisfice, and satisficing does not lead to further questions about behavior, because any "less than perfect" result can be written off as satisficing.

9   Testing Foraging Models

Only the behavior and ecology of real animals can determine the ultimate value of foraging models. Can the models explain existing observations? Do they predict new phenomena? Can they make quantitative and general predictions? Although we have discussed some empirical evidence in preceding chapters, here we consider these questions in more detail. First we consider *what kind* of account of the data foraging theory might provide, together with the related question of how one might "test" a foraging model. Second, we summarize the available evidence, thereby showing the *degree of success* that the prey and patch models have enjoyed in studies that have tested them. This summary also helps to answer our third question: what are some *common pitfalls* that testers of foraging models meet? Finally, we suggest some guidelines for those planning to use the models in interpreting their data; some of the points we raise may seem trivially obvious, but the frequency of errors in the literature indicates the need to spell them out.

9.1 Foraging Models and Data

WHAT ARE FORAGING MODELS SUPPOSED TO DO?

Different people have used foraging models (and optimality models generally) for different purposes. We distinguish six ways to use foraging models, although we do not contend that our list is exhaustive.

1. *To Ask How Good Organisms Are at Doing Their Jobs.* In other words, how well adapted are they? McFarland (1977) outlined this research program, in which one compares the animal's actual decision rules with a range of alternatives. If the actual rules maximize fitness, then McFarland would claim that the animal is well adapted. This program has yet to be carried out.

2. *To Ask What Animals Are "Designed" To Do.* This use of foraging models emphasizes currencies; it asks which currency gives the best account of behavior, given that the constraints are well understood. This kind of research might eventually show how differences in currency are

correlated with differences in ecology (e.g. maximizing net energy gain may explain the behavior of small animals with high metabolic rates, but maximizing efficiency may explain the behavior of animals that operate within a fixed energy budget). Caraco's analysis (Caraco et al. 1980b, Chapter 6) of risk-sensitive choice shows, for example, how differences in currency might be related to ecological conditions.

3. *To Analyze Behavioral Mechanisms.* If the currency is assumed to be well understood, foraging models could be used to determine which constraint assumptions best account for behavior. Since constraints reflect the mechanisms controlling behavior, this approach might increase our understanding of mechanisms. For example, Cheverton et al. (1985) analyzed bumblebee movements using this approach; assuming that the bees maximize the rate of gain, Cheverton et al. used bees' "errors" in performance to make deductions about the mechanisms that control movement between flowers.

4. *To Simultaneously Analyze What Animals Are Designed To Do and How They Do It.* When using foraging models in this way, workers acknowledge uncertainty about both constraints and currency. The problem inherent in this approach is that there are two "unknowns" (constraints and currencies) to play with (Chapter 8), and usually one must assume that either constraints or currencies are understood (Cheverton et al. 1985).

5. *To Serve as a General Background Against Which To Organize Observations about Individual Behavior.* Many field studies, although not explicitly aiming to "test" foraging models, use the general ideas of foraging to organize data and ideas. A typical example might be using the prey model to account for seasonal changes in diet breadth (Schluter 1981). This approach is probably used more often than the others.

6. *To Serve as a Tool for Understanding the Organization of Communities or Populations.* This enterprise has received little emphasis in this book, although it was the aim of the original papers in foraging theory (i.e. Emlen 1966, MacArthur and Pianka 1966). We believe it may be best to establish how well the theory accounts for individual behavior in well-defined situations, and then proceed to more complex problems.

POSSIBLE OUTCOMES OF TESTS OF FORAGING THEORY

Even the most skeptical workers would admit that current foraging models can qualitatively account for what some foragers do some of the time. At the other extreme, even the most zealous advocates of current models would admit that they do not account for all foraging behavior in quantitative detail. It is surely uncontroversial to conclude that reality lies somewhere between these two extremes! Obviously, the advocates of foraging theory would prefer to find that foraging theory gives quantitative and general accounts of foraging behavior, since an imprecise and parochial model is less desirable than one that makes precise and general statements about the things it purports to explain.

There is, however, a trade-off: the greater the precision, the less the generality, and vice versa (Holling 1966); foraging models aim for generality rather than for complete precision. The most precise models in biology are usually purely descriptive: they are based on observed relationships or processes, but foraging models are (to some extent) predictive and explanatory because they are derived (at least in part) from *a priori* considerations. For example, a purely descriptive model might be based on field estimates of the population processes of mortality, fecundity, dispersal, and so on. Such a model emphasizes accurate descriptions of a particular population, and the descriptive model used to manage a duck population may not work for an anchovy population. In contrast the marginal-value theorem might apply equally well to fish, birds, and insects.

The basic foraging models do not predict between- and within-individual variability in behavior, and this limits the accuracy of their predictions (an exception is the "ideal free" model of Fretwell 1972 that can account for variability in habitat choice on a population level; see also Parker 1978, Milinski 1979, Harper 1982, Godin and Keenleyside 1984). This is not to say that foraging theory cannot explain variability: one approach models variability as information acquisition (Chapter 4—subjecting variability itself to an optimality analysis); another approach treats variability as a constraint (Houston and McNamara 1985).

9.2 Testing Foraging Models

Much of the literature on foraging theory in the last ten years has tried to test the basic foraging models' predictions. Ideally, a test should ensure that the model's assumptions are met, and the observer should directly measure the predicted quantities (such as patch residence time, proportion of different prey types attacked upon encounter, or size of items brought to the nest). How closely should the predictions agree with the results? The discussion in the previous paragraph indicates that we should not expect *exact* correspondence, so determining how much disparity is "acceptable" becomes a matter of judgment. Standard statistical techniques, of course, provide one yardstick by which the agreement between predicted and observed outcomes may be judged (see Box 2.3 on the statistical treatment of prediction of "all or nothing" choices).

Predictions versus assumptions. In principle any model can be investigated by examining its assumptions, its predictions, or both. Foraging models make two types of assumptions. Some assumptions are part of the general background (e.g. the assumptions that net rate of energy gain is related to fitness, and that natural selection optimizes design), but other assumptions are specific to the model being considered (e.g. the incompatibility of search and handling, within-patch resource depression, or sequential encounters). Assumptions that constitute the general background usually are not tested directly, and indeed they may not be directly testable. Instead, they gain or lose credibility with the successes or failures of models based on them: when the predictions of a foraging model agree with the data, the model's assumptions—including those we call "general background assumptions"—are vindicated.

"Model-specific" assumptions can usually be tested directly, by observation or by appropriate experimental procedures. In many foraging studies the observer does not know which assumptions, if any, are met (see Table 9.1); how should these tests be assessed? A quantitative agreement between predicted and observed outcomes may partially justify the (untested) assumptions (although caution is necessary—see section 9.5). However, since the fit is rarely exact, one must know whether specific assumptions are met to determine what may have caused the discrepancy between predicted and observed outcomes. In many of the studies in Table 9.1 results that apparently run counter to the foraging models' predictions may have done so because the model's assumptions were not met. Thus they are not the strong refutations needed to guide further research.

Alternative hypotheses. Most studies that have tested foraging models have considered only one explicit alternative hypothesis, namely, a null hypothesis of random choice. For example, a study of how patch residence time is related to travel time may conclude that the two are positively related, and that the relationship between them is not random. A quantitative test of the relationship between patch residence time and travel time (e.g. Cowie 1977), in which a particular form of the relationship is predicted, implicitly tests alternative hypotheses such as linear and hyperbolic relationships.

Most people think of alternative hypotheses as different models incorporating different constraints or currencies, and not as different statistical relationships between variables. Some studies explicitly set out to compare currencies (e.g. Caraco et al. 1980b, Kacelnik 1984, Schmid-Hempel et al. 1985, Stephens et al. 1986), and this approach should be more widely adopted.

Table 9.1

Summary of tests of classical foraging models and their derivatives

	Codes
Prey models	
Where?	L = Laboratory F = Field
How?	E = Experiment O = Observation
Predictions tested	A = Preference for more profitable prey
	B = Increased selectivity at higher encounter rates
	C = Selectivity independent of abundance of low-ranking prey
	D = Quantitative estimation of threshold for dropping items from diet
Assumptions	Exc. = Exclusivity of search and handling
	Sec. = Sequential or simultaneous encounters
	Ran. = Random encounters
	Inv. = Involvement time (handling and recognition)
	(In each case ✓ indicates that assumption is probably correct for model used, ? indicates it is not known whether assumption is correct, and ✕ indicates assumption is incorrect.)
Results	1 = Quantitative agreement with model
	2 = Quantitative agreement but partial preferences
	3 = Partially or qualitatively consistent with model
	4 = Inconsistent with model
Patch (marginal-value) models and Central-place foraging models (multiple-prey loaders)	
Codes as for prey models except:	
Predictions tested	A = More persistent in each patch when environment as a whole poor or when travel time longer
	B = Patches all reduced to similar marginal value
	C = More resources extracted from better patches
	D = Quantitative tests
Assumptions	I = Patch quality recognizable
	G = Gain function known
Central-place foraging models (single-prey loaders)	
Codes as for prey models except:	
Predictions tested	A = Bigger prey brought from greater distances
	B = Quantitative estimation of distance for dropping prey size classes from diet
Assumptions	V = Availability of sizes similar at all sites

Table 9.1 (Continued)

Prey models

Reference		Where? L/F	How? E/O	What? Forager	What? Prey	Predictions tested A	B	C	D	Assumptions Exc.	Seq.	Ran.	Inv.	Results	Comments
Allen	1983	L	E	Asterias	bivalves	✓	✓			✓	?	?	✓	3	
Anderson	1984	L	E	Micropterus	guppies, damselflies	✓	✓			✓	?	?	✓	3	Fish more selective when apparent prey density higher (less cover in tank)
Barnard and Brown	1981	L	E	Sorex	Tenebrio	✓	✓	✓		✓	✓	✓	✓	3–4	Preference based on size, not e/h
Barnard and Stephens	1981	F	O	Vanellus	Lumbricus	✓				✓	✓	✓	✓	3	Prefer more profitable flowers
Bell et al.	1984	F	O	Hymenoptera	Impatiens	✓				✓	✓	?	✓	3	Prefer more profitable flowers
Belovsky	1978	F	O	Alces	vegetation				✓	NA	NA	NA	NA	1	Nutrient constraint model agrees with observed diet
Belovsky	1981	F	O	Alces	vegetation				✓	NA	NA	NA	NA	3–4	Choice of some species within habitat not as predicted; others taken if above nutrient threshold
Davidson	1978	F	O	Pogonomyrmex	barley seeds	✓	✓			✓	✓	✓	?	3	Increasing selectivity at high density
Davies	1977a	F	O	Motacilla	insects	✓	✓	✓		✓	?	?	?	3	Increasing selectivity at high density. No effect of abundance of low-ranking prey
Davies	1977b	F	O	Muscicapa	insects	✓	✓	✓		✓	?	?	?	3	
Draulans	1982	F	E/O	Aythya	Dreissena	✓	✓			✓	?	?	?	3–4	Consistent with model only if assume dive time is limited
Ebersole and Wilson	1980	L	E	Peromyscus	seeds	✓				?	?	?	?	?	e/h not measured: results therefore inconclusive
Elner and Hughes	1978	L	E	Carcinus	Mytilus	✓	✓			✓	✓	✓	✓	3	Consistent with model including recognition time
Emlen and Emlen	1975	L	E	Mus	seeds	✓	✓	✓		✓	?	?	?	3	See Rechten et al. (1981)
Erichsen et al.	1980	L	E	Parus	Tenebrio	✓	✓	✓		✓	✓	✓	✓	2	
Eringe	1981	L/F	O	Mustela (♂)	voles	✓	✓	✓	✓	✓	?	?	?	3–4	♂ prefers more profitable species; ♀ does not. Prey distribution not known
Furnass	1979	L	E	Perca	Crustacea	✓				✓	?	?	?	?	e/h not measured: results therefore inconclusive
Gardner	1981	L	E	Lepomis	Daphnia	✓			✓	✓	✓	✓	✓	3	Rejection of small prey not due to apparent size, but see Butler and Bence (1984), Wetterer and Bishop (1985)

Reference	Year			Predator	Prey							n	Comments
Getty and Krebs	1985	L	E	*Parus*	*Musca* pupae	√	√	√	√	√		3	Consistent with signal detection model of optimal prey choice
Gibson	1980	L	E	*Gasterosteus*	*Daphnia*	√√	√	?	?	√√		3	See Wetterer and Bishop (1985)
Gittelman	1978	L	E	*Notonecta/ Neoplea*	Corixids/ *Daphnia*	√√	?	?	√√	?		?	No increase in selectivity at higher prey densities, but prey model may be inappropriate (see Cook and Cockrell 1978, Sih 1980)
Goss-Custard	1977a	F	O	*Tringa*	polychaetes	√√	√	√	?	?		2	*Corophium* preferred to *Nereis* even though *e/h* lower; availability a possible confounding variable
Goss-Custard	1977b	F	O	*Tringa*	*Corophium*	√√	√	√	?	?		4	
Hames and Vickers	1982	F	O	*Homo sapiens*	various	√	√	?	?	?		3–?	Handling time not measured: significance not clear
Hawkes et al.	1982	F	O	*Homo sapiens*	various	√√	?	?	?	√		3	*e/h* not measured
Horn	1983	F	O	intertidal fish	algae	√√	?	?	?	√		?	
Houston et al.	1980	L	E	*Parus*	*Tenebrio*	√√	√	√	√√	√		2–3	See Rechten et al. (1981). *Ad hoc* modification of model for one bird (n = 5)
Hughes and Elner	1979	L	E	*Carcinus*	*Nucella*	√√	√√	√√	√√	√√		4	Do not prefer most profitable size
Hughes and Seed	1980	L	E	*Callinectes*	*Mytilus*	√√	√√	√√	√√	√√		3	Prefer smallest prey; size a possible confounding variable
Jaeger and Barnard	1981	L	E	*Plethodon*	*Drosophila*	√	√	√	√	√		3	Prefer large prey; size a possible confounding variable. Preference increases with encounter rate; some effect of encounter rate with low-ranked prey
Jaeger and Rubin	1982	L	E	*Plethodon*	*Drosophila*	√	?	?	√√	√√		3–4	Prefer large prey only if experienced
Kaufman and Collier	1981	L	E	*Rattus*	*Helianthus*	√	√	X	X	√√		3	Prefer husked seeds
Kislalioglu and Gibson	1976	L	E	*Spinachia*	crustaceans	√	√	?	?	?		3	See Rechten et al. (1981)
Krebs et al.	1977	L	E	*Parus*	*Tenebrio*	√√	√	√	√	√		2	*e/h* not measured: relevance not clear
Lacher et al.	1982	L	E	*Kerodon*	leaves	√	?	?	?	√		?	
Lea	1979	L	E	*Columba*	mixed grain	√	√	√	√	√		2–4	Main discrepancy: prefer small reward with short delay over big reward with long delay even if former has lower *e/h*
Lewis	1980	F	O	*Sciurus*	*Quercus* seeds	√		?	?	?		3	
Lewis	1982	F	O	*Sciurus*	*Carya* seeds	√		√	?	?		4	Hickory seeds preferred even though others had higher *e/h*
Lobel and Ogden	1980	F	O	*Spirasoma*	algae	√		√	?	?		3–4	Exception: one toxic species avoided

(Continued)

Table 9.1 (Continued)

Reference		Where? L/F	How? E/O	What? Forager	What? Prey	Predictions tested A	B	C	D	Assumptions Exc.	Seq.	Ran.	Inv.	Results	Comments
Marden and Waddington	1981	F	E	Apis	artificial flowers	✓				✓	✓	?	✓	3	
Meire and Ervynck	1985	F	O	Ostralegus	molluscs	✓				✓	✓	✓	✓	3	Prefer leaves with high protein/fiber. Time variables not measured
Milton	1979	F	O	Howler monkey	leaves	✓		✓		✓	✓	?	?	3–?	
Mittelbach	1981	L/F	E/O	Lepomis	Daphnia	✓			✓	✓	?	✓	✓	2	Apparent size a confounding variable; encounter courld be simultaneous (see Wetterer and Bishop 1985)
Moermond and Denslow	1983	L	E	Frugivorous birds	Fruit	✓				✓	X	X	✓	?	Choice tests: large, ripe or accessible fruit preferred
Montgomerie et al.	1984	L	E	Hummingbirds	artificial flowers	✓				✓	✓	✓	✓	2	Choice maximizes net energy/volume, not net rate of intake
Ohguchi and Aoki	1983	F	E	Apis	food/water	✓			✓	✓	?	?	✓	2–4	Maximize energy gain when water not in short supply
Palmer	1979	L	E	Thais	invertebrates	✓	✓	✓	✓	✓	?	✓	✓	2–3	
Pastorok	1980	L	E	Chaoborus	Diaptomus Daphnia	✓	✓			✓	✓	?	?	3	More selective when less hungry and at higher encounter rates
Pulliam	1980	F	O	Spizella	seeds	✓	✓			✓	✓	?	?	3–4	Some exceptions to preference based on profitability. Spatial distribution not known
Rapport	1980	L	E	Stentor	Protists	NA	NA	NA		NA	NA	NA	NA	3	Results consistent with complementary resources model
Rechten et al.	1983	L	E	Parus	Tenebrio	✓			✓	✓	✓	✓	✓	3	Partial preference explained by discrimination errors. Allowing for this, birds too choosey
Reichman	1977	F	O	desert rodents	seeds	✓				?	?	?	?	?	No measures of e/h, spatial distribution of seeds, etc.
Richards	1983	L	E	Thinopinus	amphipods	NA	NA	NA	NA	?	?	?	?	3	Less selection near satiation, consistent with prey model
Ringler	1979	L	E	Salmo	invertebrates	✓				?	✓	✓	?	3	Prefer more profitable prey when experienced; increased selectivity at high density

Author	Year	L/F	E/O	Predator	Prey						No.	Comments
Robertson and Lucas	1983	L/F	E/O	*Allorchestes*	algae	✓		?	?	?	3	Amphipods prefer the algal species that give the highest growth and reproduction
Snyderman	1983a 1983b	L	E	*Columba*	mixed grain	✓	✓	✓	✓	✓	1–4	Quantitative agreement except for short delay (handling) preferred to long (see Lea 1979)
Stein	1977	L/F	E/O	*Micropterus*	*Oreonectes*	✓	✓	✓	?	✓	3–4	Prefer profitable sizes on one substrate but not on another
Stein et al.	1984	L	E	*Lepomis*	snails	✓	✓	✓	✓	✓	3–4	Choose genera of snails with highest *e/h*, but *e/h* does not account for within-species choice of size classes
Sutherland	1932	F	O	*Haemotopus*	*Cerasioderma*	✓	✓	✓	?	?	3	*Cerapteryx* has lower *e/h* but is preferred
Tinbergen	1931	F	E/O	*Sturnus*	*Cerapteryx/ Tipula*	✓	✓	✓	?	✓	4	
Thompson	1983	F	O	*Pluvialis*	worms	✓	✓	✓	✓	✓	3	Select profitable sizes allowing for effects of kleptoparasitism
Thompson and Barnard	1984	F	O	Plovers (2 species)	worms	✓	✓	✓	✓	✓	3	
Turner	1982	F	O	*Hirundo*	insects	✓	✓	?	?	?	3–4	Percent of small insects brought to young correlates with relative abundance
Vadas	1977	L/F	E/O	*Strongylocentrotus*	algae	✓	✓	✓	?	?	3	Prefer species with highest assimilation efficiency. No direct test of prey model
Vickery	1984	F	E	Rodents	fruit	✓		?	×	✓	?	Inappropriate application of prey model to choice tests
Waddington and Holden	1979	L	E	*Apis*	artificial flowers	✓	✓	✓	✓	✓	2–3	Consistent with simultaneous encounter model. Individual variability
Wells and Wells	1983	F	E	*Apis*	artificial flowers	✓	✓	?	?	×	4	Do not choose profitable prey, but *e/h* not measured
Werner and Hall	1974	L	E	*Lepomis*	*Daphnia*	✓	✓	?	✓	✓	2–3	Encounters may not be sequential; apparent size a confounding variable (see Wetterer and Bishop 1985)
Wetterer and Bishop	1985	L	E	*Culaea*	*Daphnia*	✓	✓	✓	✓	✓	4	Fish choose *Daphnia* on apparent size, not as predicted by optimal foraging theory
Winterhalder	1981	F	O	*Homo sapiens*	varicus	✓	✓	✓	?	✓	3	More selective at higher density
Zach and Falls	1978	L	E	*Seiurus*	insects	✓	✓	✓	✓	✓	3	Experiment 3 in this paper is the one that is relevant to the prey model

(*Continued*)

Table 9.1 (Continued)

Patch (marginal-value) models

Reference		Where? L/F	How? E/O	Forager (What?)	Prey	A	B	C	D	Exc.	Seq.	Ran.	I	G	Results	Comments
Alphen and Galis	1982	L	E	Asobara	Drosophila				✓	✓	?	?	✓	?	?	Only one patch at a time presented; results not clear
Best and Bierzychudek	1982	F	O	Bombus	Digitalis					✓	✓	?	✓	✓	1–3	Discrepancy is that bees sometimes skip flowers
Bond	1980	L	E	Chrysopa	Acyrthosiphon	✓				?	?	?	?	?	3	More persistent area-restricted search when hungry; relevance to patch model not clear
Cook and Cockrell	1978	L	E	Adalia larvae	Acyrthosiphon	✓				✓	✓	✓	?	✓	3	Prey treated as a patch (Notonecta results in same paper: see Chap. 4)
Corbet et al.	1981	F	O	Vespa/Bombus	Scrophularia/Linaria	✓				✓	?	?	?	✓	4	Exploit patches independent of resource depression
Cowie	1977	L	E	Parus	Tenebrio larvae	✓	✓			✓	✓	✓	✓	✓	1	Averaged data fit the model
Formanowicz	1984	L	E	Dytiscus	tadpoles	✓	✓			×	?	?	✓	?	3?	See Chap. 4 for problem of overlapping encounters
Giller	1980	L	E	Notonecta	Aedes	✓				×	×	?	✓	✓	3?	Possible problem of overlapping encounters (see Chap. 4)
Griffith	1982	L	E	Macroleon	ants	✓				×	×	?	✓	?	3?	See comments on Giller (1980)
Harting and Plowright	1979	L	E	Bombus	nectar	✓				✓	✓	?	✓	?	3?	More persistence with longer travel time, but gain function not measured
Hassell	1980	F	O	Cyzenis	Operophtera	✓		✓		✓	✓	✓	✓	?	3	
Haynes and Mesler	1984	F	O	Bombus	Lupinus pollen	✓				✓	✓	?	✓	?	4	See comments on Waddington and Heinrich (1979)
Heads and Lawton	1983	F	O	Chrysocharis	Phytomyza	✓		✓		✓	✓	?	✓	?	3	
Hodges	1984	F	E	Bombus	Delphinium	✓		✓		✓	✓	✓	✓	✓	3	More flowers visited per inflorescence when travel time long
Hodges and Wolf	1981	F	O	Bombus	Delphinium	✓	✓	✓		✓	✓	✓	?	✓	1–3	More nectar left behind in rich sites
Hubbard and Cook	1978	L	E	Nemeritis	Ephestia	✓	✓	✓		✓	?	?	✓	?	1–3	Too many prey taken from poor patches
Krebs et al.	1974	L	E	Parus	Tenebrio larvae	✓	✓			✓	✓	✓	?	?	3–?	See McNair (1982)
Larkin	1981	L	E	Barbary dove	grain	✓			✓	✓	✓	✓	✓	✓	1	Progressive interval schedule to simulate depression

Author	Year			Organism	Resource								Comments
Lewis	1980	F	O	*Sciurus*	*Quercus*	✓	✓	✓	?	?	?	?	⎫ Not clear that there is resource
Mellgren	1982	L	E	*Rattus*	pellets	✓	✓	✓	✓	?	?	?	⎬ depression
Mellgren et al.	1984	L	E	*Rattus*	pellets	✓	✓	✓	✓	?	?	?	⎭
Munger	1984	F	O	*Phynosoma*	*Pogonomyrmex*	✓	✓	✓	✓	✓	✓	3–1	Patches left when marginal capture rate equals habitat average. Better patches not exploited for longer
Parker	1984	F	E/O	*Hesperotettix*	*Guitierrezia*	✓		✓	?	?	?	3	Plants gradually depleted. Movement to new plant depends on inter-plant distance
Parker and Stuart	1976	F	O	*Scatophaga* ♂	*Scatophaga* ♀			✓	✓	✓	✓	1	Copula duration as predicted
Pyke	1978a	F	O	*Bombus*	*Delphinium*			✓	?	✓	✓	3	Exploit flowers on inflorescence in order of decreasing marginal value
Pyke	1978b	F	O	*Selasphorus*	*Ipomopsis*	✓	✓	✓	?	✓	✓	3	More persistent in better patches
Roitberg and Prokopy	1982	F	E	*Rhagoletis*	*Crategus* trees	✓	✓	?	?	?	✓	3	Search each tree longer when travel time longer, but observed patch time greater than predicted
Schmid-Hempel et al.	1985	F	E	*Apis*	nectar	✓		✓	✓	✓	✓	1	Observations close to prediction of efficiency maximizing
Sih	1980	L	E	*Notonecta*	*Culex*	✓		×	✓	?	✓	3?	See comments on Giller (1980)
Smith and Sweatman	1974	L	E	*Parus*	*Tenebrio* larvae	✓	✓	✓	?	?	?	?	Gain function not known
Townsend and Hildrew	1980	L/F	E	*Plectrocnemia*	Stream-crift Chironomids	✓	✓	×	✓	?	×	?	Giving-up time equal in different "habitats"; significance not clear
Waage	1979	L	E	*Nemeritis*	*Ephestia*	✓		✓	✓	✓	✓	3–4	More resource extracted from better patches; persistence not related to habitat abundance
Waddington and Heinrich	1979	L	E	*Bombus*	nectar	✓		✓	?	✓	✓	4	Not sensitive to vertical pattern of resource depression
Wasserman	in prep.	L	E	*Coiumba*	mixed grain	✓	✓	✓	✓	✓	✓	1–3	Progressive ratio to simulate resource depression; tendency to switch early
Williams	1982	L	E	*Rattus*	pellets	✓		✓	✓	✓	✓	1–3	See comments on Wasserman (in prep.)
Witham	1977	F	O	*Bombus*	*Chilopsis*	✓	✓	✓	✓	?	✓	1–3	Progressive ratio to simulate resource depression
Ydenberg	1982	L	E	*Parus*	*Musca* pupae	✓		✓	?	✓	✓	1–3	
Zimmerman	1981	F	O	*Bombus*	*Polemcnium*	✓		✓	?	?	?	3?	Slight increase in patch persistence as travel time increases, but gain curve not known

(Continued)

Table 9.1 (Continued)

Central-place foraging models (multiple-prey loaders)

Reference		Where? L/F	How? E/O	Forager	Prey	Predictions tested				Assumptions					Results	Comments
						A	B	C	D	Exc.	Seq.	Ran.	I	G		
Brooke	1981	F	O	*Oenanthe*	insects	✓				✓	✓	?	?	?	3	Load-size distance effect
Bryant and Turner	1982	F	O	*Delichon*	insects	✓			✓	✓	✓	?	?	✓	3	Load-size distance effect quantitatively not in agreement, but see Kacelnik and Houston (1984)
Carlsson and Moreno	1981	F	E	*Oenanthe*	*Tenebrio* larvae	✓				✓	✓	✓	✓	✓	3	Load-size distance effect
Giraldeau and Kramer	1982	F	E	*Tamias*	*Helianthus* seeds			✓		✓	✓	✓	✓	✓	3	Patch times too short; qualitative agreement
Kacelnik	1984	F	E	*Sturnus*	*Tenebrio* larvae	✓		✓		✓	✓	✓	✓	✓	1	Energy costs included
Kasuya	1982	L	E	*Polistes*	water	✓		✓		✓	✓	✓	✓	?	1–3	
Killeen et al.	1981	L	E	*Rattus*	pellets	✓		✓		✓	✓	?	✓	?	3?	Load-size distance effect; gain function ambiguous

Central-place foraging models (single-prey loaders)

Reference		Where? L/F	How? E/O	Forager	Prey	Predictions tested		Assumptions					Results	Comments
						A	B	Exc.	Seq.	Ran.	Inv.	V.		
Carlsson	1983	F	E	*Lanius*	*Tenebrio*	✓	✓	✓	✓	✓	✓	✓	2–4	Model supported at close and distant sites, not intermediate ones
Davidson	1978	F	E	*Pogonomyrmex*	barley seeds	✓		✓	?	?	?	✓	3	Larger seeds and narrower range of seeds from greater distances
Hartwick	1976	F	O	*Ostralegus*	intertidal invertebrates	✓		✓	?	?	?	✓	3	Difference in parental and off-spring diet. Parents' items smaller than those of young
Hegner	1982	F	O	*Merops*	insects	✓	✓	✓	✓	✓	✓	✗	2–3	More selective at greater distances
Jenkins	1980	F	O	*Castor*	trees	✓		✓	?	?	✓	?	3	More selective at greater distances
Krebs and Avery	1984	F	O	*Merops*	insects	✓	✓	✓	✓	✓	✓	✓	2–3	More selective at greater distances
Lind	1965	F	O	*Ostralegus*	molluscs	✓		✓	✓	?	?	✓	3	See comments on Hartwick (1976)
McGinley	1985	L	E	*Neotoma*	sticks	✓		✓	✓	?	?	?	3	More selective at greater distances
Royama	1970	F	O	*Parus*	insects	✓		✓	✓	?	?	?	3	See comments on Hartwick (1976)

Quantitative versus qualitative tests. In section 9.1 we suggested that foraging models should not be judged by their quantitative predictions alone, although these are clearly more powerful than qualitative predictions in two senses. First, they provide the most informative test of the explanatory abilities of the models, and of the potential validity of their assumptions; because many models may predict the same qualitative trends, so quantitative details can usually eliminate more alternative hypotheses than qualitative trends can. Second, quantitative predictions provide a more complete description of the forager's behavior (which is, after all, the model's aim). However, when quantitative predictions fail or when they cannot be made (as in most current trade-off models), but observed qualitative trends agree with the model under test, what should the investigator conclude? It is probably reasonable to conclude that the model has captured the essence of the situations it attempts to explain, although the danger that alternative hypotheses may explain the same trends is greater than for quantitative predictions. However, even qualitative results can be decisive: they can sometimes distinguish between two widely held views, in the way that demonstrations of risk sensitivity contradict average-rate maximizing.

9.3 How Well Does Foraging Theory Do?

Four recent papers have assessed the relationship between predictions and observations in foraging theory (Krebs et al. 1983, Pyke 1984, Gray 1986, Schoener 1986). Schoener, extending Pyke, lists five attitudes toward foraging theory that are found in the ecological literature: (1) it is trivial or tautological and "unscientific" because *post hoc* rationalization can explain any discrepancy between theory and prediction (e.g. Ghiselin 1983; see Chapter 10 for further comments); (2) it is too simple to work in the field, although it may work in laboratory experiments (Zach and Smith 1981); (3) it is so well established that no further tests are needed; (4) it is still too early to judge how well it accounts for the data; and (5) it has often been verified, and this lays the groundwork for many future developments. To these attitudes, Gray adds the view that foraging theory is largely at odds with the published data. Table 9.1 presents a modified and updated version of the table in Krebs et al., summarizing the evidence related to some of the basic models.

Table 9.2 summarizes the main conclusions that can be extracted from Table 9.1. First, most of the evidence in the literature (64%) tests qualitative rather than quantitative predictions. Second, 71% of papers report either

Table 9.2
Summary of predictions and assumptions of models in tests listed in Table 9.1[a]

	Types of predictions tested			
	Prey	Patch	CPFSPL[b]	Totals
Qualitative	51	24	6	81
Quantitative	20	21	3	44

	Assumptions met												
	Prey				Patch				CPFSPL				Overall $\bar{x}$
	✓	?	X	(%✓)	✓	?	×	(%✓)	✓	?	×	(%✓)	%✓
Exc.	58	10	0	(86)	40	0	5	(89)	9	0	0	(100)	88
Seq.	33	31	4	(49)	39	3	3	(87)	7	2	0	(78)	65
Ran.	30	34	4	(44)	24	21	0	(51)	4	5	0	(44)	47
Inv.	40	26	2	(59)	27	18	0	(57)	6	3	0	(66)	59
G	—				29	16	1	(62)	—				62
V	—				—				6	2	1	(66)	66
%	$\bar{x} = 59.5$				$\bar{x} = 69.0$				$\bar{x} = 71.0$				

	Results							
	Prey		Patch		CPFSPL		Totals	
Outcome	n^c	%	n	%	n	%	n	%
1	1.5 ⎫		9.5				11.0 ⎫	
2	8.0 ⎬ 70.0		0.0	70.0	1.5 ⎫		9.5 ⎭ 16.4	
3	40.0 ⎭		22.0		7.0 ⎭ 94.0		69.0	55.2
4	12.5		3.5		0.5		16.5	13.2
?	9.0		10.0				19.0	15.2
Total	71.0		45.0		9.0		125.0	

[a] In this and subsequent tables, results in Table 9.1 that have two entries (e.g. 3–4) are scored ½ to each outcome. Patch category includes multiple-prey loaders.
[b] CPFSPL = Central-place foraging (single-prey loaders).
[c] n = Number of studies with a given outcome.

qualitative or quantitative support for the models, and of the remainder only 13% clearly contradict the predictions. Third, the prey and patch models are about equally supported, with the exception that the zero-one rule is never completely supported. As we have discussed in Chapter 2 (Box 2.3, Stephens 1985), this may not be a serious problem. Fourth, there is considerable variation in the extent to which specific assumptions of each model are known to be met. The exclusivity assumption is nearly always known to be met (in 88% of papers), but randomness of encounters in prey and patch models and the gain function in the patch model are

Table 9.3

Status of assumptions in studies in which model is strongly
supported[a] or strongly rejected[b,c]

	Supported				Rejected			
	✓	?	×	(%✓)	✓	?	×	(%✓)
Prey	37	7	0	(84)	30	25	1	(54)
Patch	64	6	0	(91)	5	0	5	(75)
CPFSPL[d]	15	0	1	(94)				
Total	116	13	1	(89)	45	25	6	(59)

[a] In Table 9.1: categories 1, 2, 1–3, 1–4, 2–3, 2–4 (11 studies).
[b] In Table 9.1: categories 3–4, 4 (14 studies).
[c] All studies in Table 9.1 in which the outcome is 3 are excluded
from this table.
[d] CPFSPL = Central-place foraging (single-prey loaders).

seldom verified. This is an important point, because failing to verify the
assumptions renders many of the supposed tests ambiguous. Indeed, some
supposed falsifications of foraging models may be attributable to failure
to meet the assumptions. Table 9.3 shows that in studies that strongly
support the models (results 1 and 2) 89% of the specific assumptions have
been verified, but in studies that show weaker support (result 4), only 59%
of the assumptions have been met. Why do many studies fail to meet the
assumptions of the model? The most common reason is that many workers
have not set out to test the models; instead, they have applied the models
post hoc to data such as stomach contents. Another factor is that some
workers simply do not know which assumptions are involved.

Table 9.4 divides the results into field and laboratory tests. Contrary
to the claim of attitude (2) listed by Schoener (1986) and discussed above,
field tests are just as likely to support the models as laboratory tests (see
also Gray 1986). Table 9.5 summarizes the outcomes of tests that have
tried to examine quantitative predictions and of those that have looked
only at qualitative predictions. *Strong* disagreement is, if anything, less
likely in quantitative tests (10%) than in qualitative tests (19%). This may

Table 9.4

Relative success of foraging models in field and
laboratory tests

	Outcome					
	1	2	3	4	?	N
Field	6.0	3.5	39	9.0	5.5	
Laboratory	5.0	6.0	30	7.5	13.5	62

Table 9.5

Outcome of tests of prey, patch, and CPFSPL[a] models that
examine qualitative predictions and quantitative predictions[b]

	Qualitative tests	Quantitative tests
Strong disagreement (4)	12.0	4.5
Other (excluding ambiguous) (1–3)	51.0	38.5
% disagreement	19.0	10.5

[a] CPFSPL = Central-place foraging (single-prey loaders).
[b] Outcomes indicated by a ? in Table 9.1 are omitted here.

be because quantitative tests are generally those in which the models' assumptions are most carefully checked.

Based on these summaries of the evidence (Table 9.1 will undoubtedly be out of date by the time this book is published), we agree with Schoener's (1986) view that attitude (3)—success should lead to abandoning the approach—is a little perverse; attitude (4)—it is too early to judge—is unnecessarily cautious; and attitude (5)—foraging theory lays the groundwork for future developments—is the most appropriate. The basic models have accounted for the data well, and they offer the promise of future developments. Our conclusion seems at odds with Gray's (1986) claim that the data do not support the models well. Among the main reasons for this discrepancy in opinion are: (1) Gray considers the failure to observe absolute preferences (as the prey model's zero-one rule apparently predicts) as evidence against the prey model, while we take the view that a statistically estimated threshold (which predicts partial preferences) is consistent with the prey model; (2) Gray considers the prey model to be *the* basic model, and he counts results that are consistent with modifications of the prey model (such as Elner and Hughes's 1978 study of how non-zero recognition times change the prey model's predictions) as not supporting the theory; and (3) Gray does not consider whether a study meets the assumptions of the model it applies. As Table 9.3 shows, this is an important factor in assessing the literature. Obviously, there is ambiguity in interpreting some studies, and we refer the skeptical reader to the original papers.

FUTURE EMPIRICAL WORK

Table 9.1 shows that, when they have been properly tested, the prey, patch, and central-place models are on the whole qualitatively supported. Foraging theory can qualitatively account for foraging decisions. Given the evidence in Table 9.1, then, what kinds of experimental work might be valuable in the future?

The first important point is that Table 9.1 includes only a few models (patch, prey, and central-place) because most empirical studies have examined these models. Data gatherers have paid little attention to certain models, in particular those dealing with information (Chapter 4) and risk (Chapter 6). Incomplete information models present an opportunity to strengthen the link between foraging and learning theories, which has just begun to develop (e.g. Lester 1984, Regelmann 1984, Shettleworth 1984). Dynamic models of the life history aspects of foraging (Chapter 7) are also little explored. In addition, totally new theoretical issues will probably arise; witness the burgeoning interest in risk-sensitive foraging—an interest that was practically nonexistent before the publication of Caraco's work six years ago.

Second, we suggest it is worth doing more tests of the basic models and their modifications. As more evidence accumulates, foraging theory can begin to concentrate on situations in which the models fail. This pattern of failure versus success may suggest new insights on the limitations and applicability of the models. Clearly, the present evidence leaves many taxonomic lacunae: most of the work to date has been done on insectivorous birds, fish, and insects, and on a few mammals. There is notably little work on herbivores of any taxon. If we look at foraging theory in the long term, we can speculate on the kinds of questions that might evolve from current theory. The studies that have tested predictions of the basic models (Table 9.1) can be viewed as attempts to validate foraging theory's approach. It would be disappointing if this was still the major focus ten years from now. Instead, we expect that the results of foraging theory should be increasingly applied to community and population ecology (their original context) and to behavioral mechanisms (e.g. of learning and perception).

9.4 Pitfalls in Testing Foraging Models

We can easily find misunderstandings of foraging theory in the literature. A commonly held view is that there is a *single* foraging model; the hypothetical comment, "Optimal foraging theory predicts that animals maximize their rate of energy gain, but for herbivores protein is more important than energy; therefore *the* theory is wrong," typifies this view. We hope that by now even the least observant reader will have noted that there is no such thing as "*the* optimal way to forage." Foraging theory is a point of view, comprising a large and continuously changing set of models. It happens that many of the basic models adopt the premise of average-rate maximizing, but this is no more an essential part of foraging theory than simple Mendelian ratios are an essential part of genetic analysis.

Beyond these general misunderstandings we can also find many examples of the wrong model being applied to a particular data set, or of models being applied without having their assumptions verified. The commonest mistakes include failure to directly observe decisions made upon encounter in tests of the prey model and failure to show patch depression (i.e. negative acceleration of the gain function) in tests of the marginal-value theorem (Table 9.2). Two "gain function" mistakes are especially common. The first mistake is to assume patch depression when the biology of the situation indicates that there should be none (e.g. suspension feeders catching stream drift—Townsend and Hildrew 1980). The second mistake is to measure the gain function without experimental control of the time spent feeding in patches (Formanowicz 1984): Whenever possible, gain functions should be measured independent of the forager's patch-leaving decision. Without independent control patches may appear to "depress," because "more difficult" patches may be visited longer, even if there is no within-patch depression.

ENERGY COSTS: AN EXAMPLE
Rather than work through all possible analytical errors, we illustrate the point with one example, energy costs. As discussed in Chapter 2, the conventional foraging models use *net* rate of energy gain as the currency to be maximized. Some tests have simply ignored costs and tested predictions based on maximization of gross energy intake (e.g. Krebs et al. 1977). Other studies have estimated the energy costs of travel, pursuit, search, and handling and subtracted them from gross energy intake (e.g. Cowie 1977, Bryant and Turner 1982, Kacelnik 1984). Figure 9.1 shows the effect of including energy costs in the marginal-value model. There are two important points. First, the effect of energy cost on the predicted residence time (or load size) depends on the *difference* in energy costs of travel and search, and not on their absolute values. If search within a patch (including pursuit and handling) costs the *same* as travel (e.g. the predator walks both when traveling and when searching), then incorporating energy costs does not alter the marginal-value theorem's patch-residence-time predictions (Fig. 9.1[A]). If travel costs *more* than search, then including costs increases the predicted patch residence time (Fig. 9.1[B], Cowie 1977). If travel costs *less* than search, as might be the case when travel involves gliding flight and search involves flapping flight (Bryant and Westerterp 1980), then including costs reduces the predicted patch residence time.

The house martin (*Delichon urbica*) is a multiple-prey loader, and Bryant and Turner (1982) used the marginal-value model (Orians and Pearson's 1979 version) to predict how the distance between the nest and the feeding

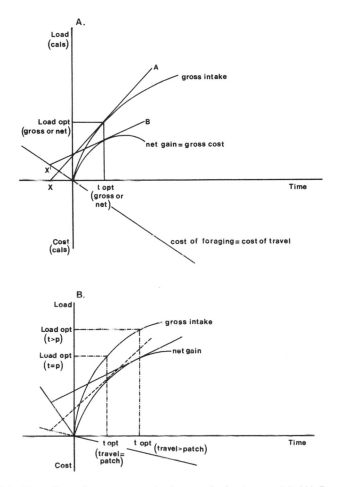

Figure 9.1 The effect of energy costs in the marginal-value model. (A) Energy costs of travel and search are the same. Gross intake is maximized by fitting a tangent from X to intersect the gross intake curve (marked A). Costs are represented by rotating the abscissa: this means that travel costs are in effect subtracted from net gain in the patch by taking the tangent from X′ instead of from X. Net intake in the patch is calculated by subtracting in-patch costs from the gross intake curve. Note that the optimal patch time is not altered when within-patch and between-patch energy costs (per time unit) are the same. (B) When travel costs more than search (represented by the steeper slope of the travel cost line), predicted load size and residence time are greater than they are when costs are equal.

site should affect the load size brought back to the nestlings. They found that although load size increased with distance in qualitative agreement with the model, *gross* energy maximization predicted loads 20 to 39% smaller than those observed. However, as Kacelnik and Houston (1984) point out, Bryant and Turner incorrectly conclude that when *net* energy costs are considered, the deviation from predictions is even greater than it is for the gross energy maximization model. This is because Bryant and Turner only subtract the costs of *within*-patch foraging from gross intake, leading them to conclude that the "net energy" model predicts a smaller load size than the gross energy model, that is, a load size even further from the observed size. Bryant and Turner have effectively assumed, since they do not correct for travel costs, that travel costs less than search; if, however, travel costs as much or more than search, the effect of correcting for energy costs would be to bring the prediction closer to the observed load size.

Bryant and Turner's paper also illustrates a second kind of confusion about energy costs. These authors calculate the predicted load size from the *net* gain curve in Figure 9.1(A), but the correct procedure is to calculate (using tangent X'B) the optimal residence time from the net gain curve, and to use this value to predict load size from the gross intake curve (see Fig. 9.1[B]). Bryant and Turner compound two errors by taking a tangent from X (instead of from X') to the net gain curve to predict optimal *gross* intake. Kacelnik and Houston (1984) discuss these points in more detail.

9.5 Sufficient Tests?

An often-cited example of a convincing test is the relationship between patch residence time (or load size for multiple-prey loading central-place foragers) and travel time (Krebs and McCleery 1984, Krebs et al. 1983). However, this evidence may not be as convincing as it seems. The marginal-value theorem predicts this relationship when the gain function is negatively accelerated. If the gain function is linear up to some maximum, then the marginal-value theorem predicts that travel time should not affect patch residence time or load size (Fig. 9.2[A]). However, Kacelnik and Cuthill (1986) have shown that central-place-foraging starlings (*Sturnus vulgaris*) take larger loads from more distant patches even when the gain function is linear (Fig. 9.2[B]).

How should this evidence be interpreted? First, it shows that the "load size-distance" effect is not necessarily related to resource depression and should not be regarded as sufficient support for the marginal-value model. Of course the better the fit, the more likely it is that the marginal-value

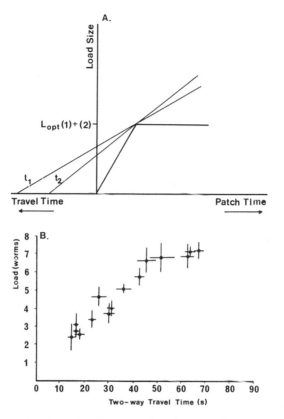

Figure 9.2 (A) With a linear loading curve there is no predicted effect of travel time on load size or patch residence time. (B) Data for starlings bringing mealworms to the nest when food was delivered at a feeding site according to a fixed interval schedule (a linear loading curve). Contrary to prediction, there is a load size-distance effect. (Data from Kacelnik and Cuthill 1986.)

theorem explains the observations. Second, the absence of environmental resource depression does not necessarily mean that the animal's gain function is linear. Suppose that costs accumulate as a function of residence time or load size; in this case a linear gross intake function might be translated into a decelerating net gain function (and see Staddon 1983). An obvious candidate for accumulating cost in a central-place forager is energy expenditure: the cost of carrying the load increases as a function of load size, so that even though gross intake is linear, the net gain curve has a diminishing slope. Third, as we pointed out in section 4.4 (see also Green 1980, Kacelnik and Cuthill 1986), patch assessment models predict increasing load size with distance even when there is no patch depression.

Finally, the starlings may be unable to recognize that the gain function is linear: they might "assume" resource depression. However, this possibility seems unlikely, since under other experimental conditions the birds can respond to non-depressing patches (Cuthill 1985). The general point is that careful experimental analyses may be essential for distinguishing between alternative interpretations. What may appear to be an adequate test of foraging theory at first glance may be less convincing when analyzed in more detail. There is, in fact, no infallible recipe for testing foraging models, any more than there is for doing research in general. However, we suggest that the following questions should be asked before testing a given foraging model.

1. *Are the Foragers Playing the Same Game as the Model?* If the model assumes successive encounters, patches with resource depression, incompatible search and handling, and so on, make sure that these conditions apply to the foragers used to test the model. This is a special danger with "off-the-shelf" models: the game assumed in the model may differ in some essential way from the game the foragers of interest play.

2. *Are the Assumptions of the Model Met?* This question extends the first one. Given that the animals are playing the right game, there may be critical assumptions that must be checked. To take a simple example, when testing Orians and Pearson's (1979) central-place foraging model, you must show that prey density does not vary systematically with distance from the central place. Table 3.1, Box 2.2, and Box 2.4 should be helpful in enabling a check of the assumptions of the classical prey and patch models and in suggesting alternative models if the assumptions are not met.

3. *Are the Right Variables Being Measured?* If the model predicts how energy, handling time, and encounter rate should affect prey choice, or how the shape of the gain curve should affect patch residence time, then establish that these variables can be measured for the system under study. If there are big measurement errors in any of them, then a sensitivity analysis can be used to determine how these errors affect the model's predictions. These points may seem obvious until one reads papers in which the prey model is "tested" without any measurements of handling time (e.g. Ebersole and Wilson 1980, Wells and Wells 1983).

4. *Is the Test Merely Consistent with the Model or Does It Rule Out Alternative Possibilities?* This is a difficult question to answer since, as with the "load size-distance" effect discussed earlier, an apparently convincing test may become less convincing when new evidence emerges. There are, however, some obviously inconclusive observations. The prey model predicts under certain conditions that there should be no selectivity

of prey. Although an observation of no selectivity (e.g. Gill and Wolf 1975) is consistent with the prey model, it is also consistent with a null hypothesis of random choice (see also Aronson and Givinish 1983 for a discussion of a similar point). Tests of bumblebee patch leaving provide a more subtle example. Pyke (1978a) and Hodges (1981) assume that a rate-maximizing nectar feeder should leave the current inflorescence when the expected gain from the next flower is lower than the expected gain from moving to a new inflorescence. Although this behavior is consistent with rate maximizing, it does not exclude the possibility that leaving before this point would give a higher rate of intake.

9.6 Summary

Ultimately, foraging theory must account for observed foraging behavior. Foraging theory seeks general explanations, rather than precise but parochial descriptions. To ensure strong inferences, the observer should check as many of the model's assumptions as possible, and the observer should consider plausible and explicit alternative hypotheses. Quantitative tests eliminate more alternative explanations than qualitative tests do.

A summary of empirical evidence to date shows that the current models do well. We argue that more tests of current models are needed to evaluate the pattern of failure; this pattern might suggest new approaches. Moreover, major aspects of foraging theory, for example, information and risk, have been the subject of little empirical work to date. Many testers of foraging models have made serious mistakes; we discuss the example of including energy costs in the models, and we point out common mistakes that can easily be avoided.

The need to weigh alternative interpretations carefully is illustrated by the persistence of the relationship between patch residence time and inter-patch search time, which sometimes occurs even when it should not! We outline four important questions that should be asked before undertaking a test of a foraging model.

10

Optimization Models in Behavioral Ecology: The Spandrel Meets Its Arch-Rival

Ecologists have developed a powerful quantitative theory, called optimal foraging strategy, for studying patterns of exploitation in nature.—Gould, *The Mismeasure of Man*

It is simply factually incorrect to describe evolution as always being an adaptive or optimizing process; . . . this view . . . ignores the body of knowledge built up by evolutionary genetics.— Lewontin, "Elementary Errors about Evolution"

10.1 Introduction

A distinguished colleague of ours once remarked, "If I give a seminar on "Foraging Theory," I get the usual mixture of critical interest and enthusiasm, but if I give the same talk and call it "Optimal Foraging Theory," I am much more likely to get a negative response." Optimization theory is controversial. A recent exchange in *The Behavioral and Brain Sciences* (Vol. 6, No. 3, 1983) is more reminiscent of a nineteenth-century polemic of the Bishop Wilberforce variety than of late twentieth-century science, with epithets such as "rhetorical flummery" and "puritanical disapproval" flying back and forth. Some see optimization models as powerful tools in evolutionary biology and behavior (e.g. McFarland and Houston 1981, Alexander 1982, Charnov 1982, Staddon 1983); others regard them as naive, vacuous, tautological, or just plain wrong (e.g. Lewontin 1978, 1979, 1983, Gould and Lewontin 1979, Ghiselin 1983, Myers 1983).

The reader trying to work out what is going on in the literature is thrown into further confusion by finding apparently contradictory views within the writings of the same individual. S. J. Gould, for example, is apparently in sympathy with optimization models when, referring to D'Arcy Thompson's demonstration that simple mathematical models can describe various morphological features, he writes (1980), "I can identify the abstract Thompsonian forms as optimal adaptations." Similarly, Gould's (1978) view that "certain morphological, physiological and behavioral traits should be superior *a priori* as designs for living in new environments. These traits confer fitness by an engineer's criterion of good design" seems to support optimization. Yet in Gould and Lewontin's (1979) well-known "spandrels" paper, the optimization approach is scathingly parodied as being "truly Panglossian," referring to Voltaire's fictional Dr. Pangloss who viewed even the worst calamity, such as the Lisbon earthquake, as being ultimately a good thing.

Why is optimization in biology so controversial? We do not aspire to resolve all the problems of optimization modeling here, but we hope to identify the key issues in the debate.

10.2 What Is Wrong with Optimization Models?

Criticisms of optimality modeling tend to be bound up with criticisms of the "adaptationist program" in general. Since we contend that optimization models help to circumvent some of the problems facing students of adaptation, we will concentrate on the criticisms that are most relevant to optimization models. For a forceful and telling response to general criticisms of the adaptationist program, see Mayr (1982, 1983).

Testability. This criticism arises because optimality modelers adopt the following procedure. A design-constraint hypothesis is erected and compared with observations; if the observations do not support the hypothesis, either the constraint or currency (or both) assumptions are modified in a new optimality model. "Surely," the critic says, "it is unscientific to keep shoring up the cracked facade of optimality with a scaffolding of *ad hoc* modifications; instead, one should entertain alternatives, such as the trait under study being of neutral selective value and therefore not being designed for anything."

Leaving aside the philosophical question of whether Popperian refutability is the only standard that a method inquiry should meet, we suggest that there are three ways to answer the hypothetical critic's comment. First, optimality is not the hypothesis under test; instead, it is the technique used to work out the testable implications of the specific hypotheses about design and constraint. Second, *ad hoc*-ism is by no means limited to, or necessitated by, optimality models. It is a vice that can be found lurking in most subjects; this complaint should therefore be directed at the "scruples" of some experimentalists, and not at optimality models themselves. Third, many critics confuse *ad hoc*-ism with the refinement of hypotheses. The rules of the scientific method require that a hypothesis be abandoned when it is disproved, but it trivializes the scientific method to claim that all the elements used to arrive at the hypothesis must also be abandoned. Suppose that in order to predict a stone's terminal velocity we combined the inverse square law of gravity with some claim about the way resistance should impede the movement of a falling stone. According to the most extreme opponents of *ad hoc* arguments, if our prediction failed, then we could never again call upon the inverse square law of gravity to formulate a new hypothesis. Unfortunately, nature has not arranged a tidy one-to-one relationship between explanatory ideas and phenomena.

Optimality models probably attract the "refutability" criticism because they are more overt and specific in formulating testable hypotheses than are most other methods in evolutionary biology. Even if we accepted the critic's proposal of giving up the optimality premise in favor of, say, selective neutrality, would we be better off? No, because if anything is untestable it is the hypothesis that a trait has absolutely no selective value (Dawkins 1983, Mayr 1983).

Choosing a strategy set. The "testability" criticism does raise a more profound problem. Optimality models are composite hypotheses: they are composed of assumptions about decision, constraint, and currency (see Chapter 1). Thus when an optimality model fails, the experimenter does not always know which elements of the model are wrong. Students of adaptation may be most interested in making inferences about the usefulness (or currency) of the trait under study, but the strength of inferences about currency depends on the experimenter's confidence in the assumptions about constraints. Can the constraints on design be identified, independent of showing that a certain set of constraint assumptions adequately describes the observed trait? We try to answer this question in the next paragraphs.

Oster and Wilson (1978) summarize the problem in this way: "The essentially innovative nature of the evolutionary process precludes an exhaustive list of strategies." This view appears to be incompatible with the requirements of optimality models, since the strategy set cannot be specified *a priori*, but the behavioral ecologist's plight is not as bleak as Oster and Wilson imply. The constraints on behavior depend on the animal's morphological and behavioral equipment—whether it can hold more than one piece of food at a time in its jaws, and so on (Chapters 1 and 8). A thorough knowledge of the mechanisms controlling behavior may provide an *a priori* way of specifying the constraints on performance.

This view suggests a distinction between the problems facing a behavioral ecologist and those facing a morphologist or physiologist using optimality models. The behavioral ecologist takes the animal's morphological and physiological equipment as constraints—for example, if bird X has a bill length of 5 centimeters, it cannot eat a particular size class of fish in less than 10 seconds—and studies the design of behavior within these limitations. The morphologist or physiologist has the more difficult task of evaluating the design of traits that the behavioral ecologist treats as constraints. The constraints on beak length are much more difficult to identify than the consequences of beak length for behavior. It might be said that the behavioral ecologist is simply begging the question by treating the difficult bits as constraints; if one's aim is to understand morphological and physiological evolution, then this comment is justified, but, given their more modest aim, behavioral ecologists may often be able to identify strategy sets *a priori*.

Independence of design features. Optimization models usually concern themselves with one or, at most, a small number of design features at a time. For example, the foraging models we described in Chapter 2 analyze the rules for maximizing rate of food intake, but they ignore other potential design criteria, such as finding mates, keeping dry, and getting enough to drink. In neglecting other design features, the models implicitly assume that different aspects of design are effectively independent, that maximizing rate of intake, for example, does not interfere with finding a mate. This assumption is often wrong, and in Chapter 5 we presented techniques for studying the problem of trade-offs between two or more activities (e.g. predation risk and feeding). Trade-off models push the independence assumption back a little, but they are at best a partial remedy, because they deal with only part of the animal's behavioral repertoire. Even if it were possible to build an optimality model that encompassed all aspects of design, we would, as Lewontin (1978) puts it, "be left in the hopeless position of seeing the whole organism as being adapted to the whole environment."

The student of design is in a bind. On the one hand there is the feasible piecemeal approach that makes the unrealistic assumption of independence of design, and on the other hand there is the immensely difficult holistic approach which, even if completed, may yield only a trivial conclusion. There does not seem to us to be a way of resolving this dilemma by abstract discussion. The piecemeal approach, in spite of its inherent flaw, offers a way to start and after all does not seem to do too poorly in practice: some of the time the assumption of independence is a close enough approximation to provide a working base (Lewontin 1984).

Ignoring genetic mechanisms. Although it is not the job of optimization models to explain the genetic mechanism of evolutionary change, the design features we evaluate with the models are the products of evolutionary change. Does this mean that we can only do optimality modeling if we include some genetics? The implication of Lewontin's (1978) comment,

"Optimality arguments dispense with the tedious necessity of knowing anything concrete about the genetic basis of evolution," is clear: he thinks optimization modelers should learn some genetics. What is so special about genetics? Lewontin forgets that population geneticists frequently dispense with the tedious necessity of knowing anything about the usefulness of traits or, for that matter, about development. None of the subdisciplines of evolutionary biology can pretend that it completely solves the problems of the science. Students of adaptation sometimes find it useful to smooth over the problems of genetics, just as students of population genetics find it useful to suppose that the cause of a selection coefficient is a problem for a functional morphologist (see the Preface). We can gain insight into the operation of one aspect of living organisms without knowing about all the other aspects. The optimization modeler is in the position of a person comparing design and performance of two makes of camera without knowing everything about the laws of optics or the chemistry of film processing. Perfectly valid judgments can be made about the relative performance of each design without a knowledge of exactly how cameras or films are made. No one would deny that more knowledge would allow a more subtle analysis, but complete knowledge is not essential to make a start.

Ignoring historical origins and the question of by-products. A problem in the study of design is that the traits we see today may have arisen for one purpose and currently serve another (Gould and Vrba 1982). For example, the feathers of birds might have originated as an insulation mechanism but currently are part of the flight mechanism; the unfused sutures of the skull of the human neonate appear to be a beautiful adaptation for squeezing through the birth canal, but this cannot have been the evolutionary origin of the feature, since it is also found in egg-laying birds and reptiles. It should be apparent that problems such as these, although important for the person trying to trace the historical origin of adaptations, are less of a problem for the student of the current utility of traits, which we have argued is the domain of optimization modeling.

There is a closely related issue: traits can have multiple consequences, some of which are genuine functions (meaning that variation in the consequences are subject to natural selection) and others of which are incidental by-products. Even the most ardent adaptationist would not propose an adaptive explanation for the color of a plant's roots and underground storage organs. Carrots are orange because of the compounds they store, not because orange-ness is inherently advantageous.

The function-versus-incidental-consequence debate reached its zenith before optimization modeling emerged as a tool in behavioral ecology

(Curio 1973, Krebs 1973b, Hinde 1975). Behavioral ecologists once asked, "Is the function of territorial defense to secure food or to space out against predators?" Optimization models allow contemporary behavioral ecologists to ask, "Are the observed details of territoriality consistent with a hypothesis of maximizing food returns, a hypothesis of minimizing predation risk, a compromise between the two, or none of these?" Optimization models help to free behavioral ecologists from the simple-minded approach of deciding between functions.

Frequency dependence. It would not be necessary to include this brief section were it not for the mistaken view that optimization models have been superseded, or shown to be wrong, by the development of game theory (Luria et al. 1981, Vaughan and Herrnstein 1986). This cannot be true since game theoretic models *are* optimization models (Vagner 1974). When the benefit obtained by the animal for adopting a particular strategy (say, "take the large prey only") is frequency-dependent (because it depends on what others in the population are doing), then the optimal solution is found using game theoretic techniques (Maynard Smith 1982). As Chapter 1 pointed out, game theoretic solutions contain the three elements of optimization models: decision, constraints, and currency. They differ from the models in this book only in the choice principle they use: game theory uses stability where we have used maximization or minimization.

In the foraging problems we have discussed, pay-offs are not frequency-dependent, so we use simple maximization or minimization techniques. In game theoretic optimization models the optimal solution is sometimes not a single strategy but a stable equilibrium mixture, so the occurrence of multiple equilibria is by no means incompatible with the notion of optimality.

Another point to note is that the evolutionarily stable state in a game theoretic model may be one in which the average fitness of individuals is lower than it would be in some other state. For example, in Maynard Smith's (1972) "hawk-dove" game, mean fitness would be higher in a population of pure doves than in the stable mixture of hawks and doves. This observation is not incompatible with the statement that evolutionarily stable states are optima—they are optima in the sense of being the best of the available strategies under the conditions of the model.

Asking the wrong question: extreme atomism. The most famous recent critique of optimality and of "adaptationism" in general is Gould and Lewontin's "spandrels" paper (1979). Many of their criticisms are more directly applicable to adaptive interpretations by paleontologists than to the more experimental approach adopted by behavioral ecologists, but the most general message of their paper is that adaptationists are as capable as anyone else of asking the wrong question.

The spandrels example makes the point clearly. Gould and Lewontin claim that an adaptationist would look at a series of arches and ask, "What is the function, or adaptive significance, of the spandrels?" (the triangular space between two arches and the structure the arches support), and even, "What is the adaptive significance of the decorations on the spandrels?" Their point is that the spandrels as such have no function— they are simply a consequence of the "Bauplan" of a structure supported by arches. The conclusion of this parable is not that adaptationism is inherently flawed, but that the question posed by Gould and Lewontin's adaptationist is wrong. If instead the question had been, "Why use arches to support a structure?" there would have been a sensible engineering answer. Mayr (1982) characterizes this problem as "sailing a perilous course between pseudo-explanatory reductionist atomism and stultifying non-explanatory holism."

Ironically, the story of the spandrels makes a case for the importance of adaptive arguments. Gould and Lewontin know that the spandrel is a by-product only because they have good engineering reasons to believe that the arch is the principal design element of the structure. Biological spandrels—including by-products and phylogenetic, ontogenetic, and genetic constraints—are much more difficult to recognize than architectural spandrels. Even if they serve no other purpose, well-formulated design models are needed to identify constraints: without a design hypothesis there would be no basis for postulating any kind of constraint!

10.3 Optimization and Newton's Second Coming

Many evolutionary biologists want their subject to be more like classical physics: they long for its precision, its lawfulness, its internal consistency, its generality, and its parallel powers of explanation and prediction. Philosophers (e.g. Scriven 1959) poke fun at this wishful thinking by calling it the "myth of Newton's second coming." An increasing number of philosophers see evolutionary biology as a subject that is not, and cannot be, like classical physics (Scriven 1959, Beatty 1980), and their arguments in support of this view bear directly on the controversy surrounding optimization and adaptation.

Beatty (1980) argues that criticisms of optimization take two forms. First, critics point out that many of the same mistakes that beset other disciplines occur in optimality modeling, but these mistakes are the fault of the practitioner, not of the approach. The second form of criticism amounts to the argument that optimality models are not the "general, empirical laws of nature" that the "physics" view of science requires. Optimality models fail the "physics test" because they do not specify the range of their own validity. The models of Chapters 2 and 3 say what rate-maximizing foraging should be like, but they do not say that all animals meeting criteria X, Y, and Z will be rate maximizers, as a physical theory would. Optimality models specify types of systems, and as such they are yardsticks against which to compare nature; they are not claims about what nature must be like.

For example, flattened appendages might be "for" swimming, "for" digging, or "for" nothing. Models of ideal swimming and digging systems could be used to distinguish between these possibilities, because these models would make different predictions about how an appendage "designed for" swimming would differ, in form and use, from an appendage "designed for" digging. Moreover, formal models can treat the more subtle question of whether the appendage is a compromise between swimming and digging functions.

Scriven (1959) makes a related point: evolutionary biology is the "type specimen" of a science that can often explain but seldom predict. In the "classical physics" view of science this is a contradiction, because explanation is just the reverse of prediction: explanation relates present conditions to past conditions, and prediction relates present conditions to future conditions. Scriven believes that the parallel between explanation and prediction in classical physics is a lucky accident, and not a logical necessity. He gives the example of a fisherman who "comes in to a clinic, his face and hands black from years of ultraviolet exposure, and a growth on the back of one hand is diagnosed as a small carcinoma." The attending physician is in a strong position to *explain* that the "cause *was* excessive exposure to the sun," even though the physician would have been in a weak position to predict that this fisherman would develop skin cancer. In principle the physician might have predicted the fisherman's ailment if he or she had known about other factors, such as environmental and genetic predispositions, but Scriven's point is that the absence of this knowledge does not diminish the place of the sun as a causal factor. "The search for a *really* complete account is never-ending, but the search for causes is often *entirely* successful" (Scriven 1959).

Many criticisms of optimality modeling are aimed at its use as a predictive tool. The critic charges that "natural selection does not *always* lead to optimization," but neither does a given dose of ultraviolet radiation *always* lead to skin cancer. Optimality models can have explanatory and inferential power, even if they cannot always predict. Evolutionary biologists may have to reconcile themselves to a science that does not fit together in the tidy way they suppose physics does.

10.4 Alternatives to Optimization?

Are there alternatives to optimization modeling as ways of studying design? To simply describe the complexity and beauty of design in nature is not enough. The scientist's job is not to report that things are hard to understand, but rather to explain what is going on. Explanation inevitably involves abstraction and simplification. It involves extracting a few principles while ignoring the surrounding mess. We believe that optimization models are a fruitful way to start on this enterprise. It is true that the models we have described are mere caricatures of nature, but this is what models are meant to be. They should "convey the essence of nature with great economy of detail" (Horn 1979).

In section 8.6 we pointed out that the alternative starting point that animals are "efficient but not optimal" is too loosely defined. The statement "flies are abundant" accounts for many data, but it is not specific enough to be of much use. The same applies to the statement "animals are efficient." Optimization models, by sticking their necks out, are more readily testable and therefore more likely to lead to progress.

Finally, what of the alternative of abandoning design considerations altogether and concentrating on the study of mechanisms? There are two counter-arguments to this alternative. First, as we argued in the Preface, biologists must consider the usefulness of traits if the explanation of adaptation by natural selection is to be more than a hollow tautology. Second, even the most mechanistic biologists frequently and almost unconsciously appeal to "design" arguments. The experiments of Otto Korner (cited in Sparks 1982) in the early 1900s provide an entertaining example. Korner believed that fish could not hear. To test his hypothesis, he engaged a well-known opera singer to perform before his aquaria. He watched his fish for the signs of enthusiasm and elation that the music would surely stir in their piscine hearts. Korner deduced that fish could not hear. Korner's arguments about design were dreadfully wrong. It remained for Von Frisch to show that if fish were "given a reason" to respond to sounds, by associating sounds with food, then they could learn to respond to sounds easily. As ludicrous as Korner's experiment seems, it illustrates how silly ideas about "design" can lead to silly inferences about mechanisms and specifically about limitations on mechanisms.

10.5 Summary

Optimization models are a way of studying the products of selection, namely, the design features of organisms. By formulating design hypotheses in a quantitative and rigorous way, they help to circumvent many of the criticisms leveled at the adaptationist approach. There are many criticisms of optimization modeling, including its lack of holism and its lack of attention to phylogenetic constraints. These criticisms amount to reasons why optimization models might be wrong but not why they are bound to be wrong. Design hypotheses are essential features of most biological research, and optimization models seem to be the most explicit and powerful approach to the study of adaptation currently available.

References and Index
of Citations

Numbers in square brackets are the sections in which the citations are made.

Ahmad, S. (1983). *Herbivorous insects: Host seeking behavior and mechanisms.* Academic Press, New York. [5.5]

Alberch, P. (1982). Developmental constraints in evolutionary processes. In: J. T. Bonner (ed.), *Evolution and Development*. Dahlem Conference Report No. 20, pp. 313–332. Springer-Verlag, Berlin. [8.1]

Alexander, R. McN. (1982). *Optima for Animals*. Edward Arnold, London. [10.1]

Allen, P. L. (1983). Feeding behavior of *Asterias ruleus* on soft bottom bivalves: a study of selective predation. *J. Exp. Mar. Biol.* 70:79–80. [Table 9.1]

Allison, J. (1983). Behavioral substitutes and complements. In: R. L. Mellgren (ed.), *Animal Cognition and Behavior*, pp. 1–30. North Holland, New York. [5.1, 5.3]

Anderson, O. (1984). Optimal foraging by largemouth bass in structured environments. *Ecology* 65.851–861. [Table 9.1]

Arnold, G. W. (1981). Grazing behaviour. In: F. Morley (ed.), *Grazing Animals*. Elsevier, Amsterdam. [5.5]

Arnold, S. J. (1978). The evolution of a special class of modifiable behaviors in relation to environmental pattern. *Am. Nat.* 112:415–427. [4.1]

Aronson, R. B. and T. J. Givinish. (1983). Optimal central-place foragers: a comparison with null hypotheses. *Ecology* 64:395–399. [9.5]

Barker, L. M., M. R. Best, and M. Domjan. (1977). *Learning Mechanisms in Food Selection*. Baylor University Press, Waco, TX. [5.5]

Barnard, C. J. and C. A. J. Brown. (1981). Prey size selection and competition in the common shrew. *Behav. Ecol. Sociobiol.* 8:239–243. [8.2, Table 9.1]

Barnard, C. J. and C. A. J. Brown. (1985). Risk-sensitive foraging in common shrews (*Sorex araneus* L.). *Behav. Ecol. Sociobiol.* 16:161–164. [6.3, Table 6.1]

Barnard, C. J. and H. Stephens. (1981). Prey size selection by lapwings in lapwing/gull associations. *Behaviour* 77:1–22. [Table 9.1]

Battalio R. C., J. H. Kagel, and D. N. McDonald. (1985). Animals' choices over uncertain outcomes: some initial experimental results. *Am. Econ. Rev.* 75:596–613. [6.3, Table 6.1]

Beatty, J. (1980). Optimal-design models and the strategy of model building in evolutionary biology. *Phil. Sci.* 47:532–561. [Preface, 10.3]

Bell, G., L. Lefebvre, L-A. Giraldeau, and D. Weary. (1984). Partial preference of insects for male flowers of an annual herb. *Oecologia* 64:287–294. [3.7, Table 9.1]

Bellman, R. E. (1957). *Dynamic Programming*. Princeton University Press, Princeton, NJ. [7.1, 7.5]

Belovsky, G. E. (1978). Diet optimization in a generalist herbivore: the moose. *Theor. Popul. Biol.* 14:105–134. [Table 3.1, 5.5, Box 5.3, Table 9.1]

Belovsky, G. E. (1981). Food selection by a generalist herbivore: the moose. *Ecology* 62:1020–1030. [5.5, Box 5.3, Table 9.1].

Belovsky, G. E. (1984). Herbivore optimal foraging: a comparative test of three models. *Am. Nat.* 124:97–115. [5.5, Box 5.3]

Best, L. S. and P. Bierzychudek. (1982). Pollinator foraging on foxglove (*Digitalis pupurea*): a test of a new model. *Evolution* 36:70–79. [Preface, Table 9.1]

Blau, P. A., P. Feeny, and D. S. Robson. (1978). Allylglucosinolate and herbivorous caterpillars: a contrast in toxicity and tolerance. *Science* 200: 1296–1298. [5.5]

Bobisud, L. I. and C. J. Potratz. (1976). One-trial versus multitrial learning for a predator encountering a model-mimic system. *Am. Nat.* 110:121–128. [4.1, 4.3]

Bond, A. B. (1980). Optimal foraging in a uniform habitat: the search mechanism of the green lacewing. *Anim. Behav.* 28:10–19. [Table 9.1]

Brooke, M. de L. (1981). How an adult wheatear (*Oenanthe oenanthe*) uses its territory when feeding nestlings. *J. Anim. Ecol.* 50:683–696. [Table 9.1]

Bryant, D. and A. K. Turner. (1982). Central place foraging by swallows (Hirundinae): the question of load size. *Anim. Behav.* 30:845–856. [9.4, Table 9.1]

Bryant, D. W. and K. Westerterp. (1980). The energy budget of the house martin *Delichon urbica*. *Ardea* 68:91–102. [9.4]

Bryant, J. P. and P. J. Kuropat. (1980). Selection of winter forage by subartic browsing vertebrates: the role of plant chemistry. *Ann. Rev. Ecol. Syst.* 11:261–285. [5.5]

Butler, S. M. and J. R. Bence. (1984). A diet model for planktivores that follow density-independent rules for prey selection. *Ecology* 65:1885–1894. [Table 9.1]

Capaldi, E. J. (1966). Partial reinforcement: a hypothesis of sequential effects. *Psych. Rev.* 73:459–477. [4.3]

Caraco, T. (1980). On foraging time allocation in a stochastic environment. *Ecology* 61:119–128. [6.4, 6.5, Box 6.1]

Caraco, T. (1981). Energy budgets, risk and foraging preferences in dark-eyed juncos (*Junco hymelais*). *Behav. Ecol. Sociobiol.* 8:820–830. [6.3, Table 6.1]

Caraco, T. (1982). Aspects of risk-aversion in foraging white crowned sparrows. *Anim. Behav.* 30:719–727. [6.4]

Caraco, T. (1983). White crowned sparrows (*Zonotricha leucophrys*): foraging preferences in a risky environment. *Behav. Ecol. Sociobiol.* 12:63–69. [6.3, Table 6.1]

Caraco, T. and M. Chasin. (1984). Foraging preferences: response to reward skew. *Anim. Behav.* 32:76–85. [6.4]

Caraco, T. and S. L. Lima. (In press). Survivorship, energy budgets and foraging risk. In: M. Commons, S. J. Shettleworth, and A. Kacelnik (eds.), *Sixth Harvard Symposium on the Quantitative Analyses of Behavior*. [6.4]

Caraco, T., S. Martindale, and H. R. Pulliam. (1980a). Avian flocking in the presence of a predator. *Nature* 285:400–401. [5.4]

Caraco, T., S. Martindale, and T. S. Whitham. (1980b). An empirical demonstration of risk-sensitive foraging preferences. *Anim. Behav.* 28:820–830. [Preface, 6.3, Table 6.1, 9.1, 9.2]

Carlsson, A. (1983). Maximising energy delivery to dependent young: a field experiment with red-backed shrikes (*Lanius collurio*). *J. Anim. Ecol.* 52:697–704. [3.2, Table 9.1]

Carlsson, A. and J. Moreno. (1981). Central place foraging in the wheatear (*Oenanthe oenanthe*): an experimental test. *J. Anim. Ecol.* 50:917–924. [Table 9.1]

Charnov, E. L. (1976a). Optimal foraging: attack strategy of a mantid. *Am. Nat.* 110:141–151. [2.2, Box 2.1, 6.6]

Charnov, E. L. (1976b). Optimal foraging: the marginal value theorem. *Theor. Popul. Biol.* 9:129–136. [2.3]

Charnov, E. L. (1982). *The Theory of Sex Allocation.* Princeton University Press, Princeton, NJ. [10.1]

Charnov, E. L. and G. H. Orians. (1973). Optimal foraging: some theoretical explorations. Unpublished manuscript. [2.1, 2.2, 2.3, 3.7, Table 3.1]

Charnov, E. L., G. H. Orians, and K. Hyatt. (1976). The ecological implications of resource depression. *Am. Nat.* 110:247–259. [2.3, 4.4]

Cheverton, J., A Kacelnik, and J. R. Krebs. (1985). Optimal foraging: constraints and currencies. In: B. Holldobler and M. Lindauer (eds.), *Experimental Behavioral Ecology*, pp. 109–126. G. Fischer Verlag, Stuttgart. [1.2, 8.2, 9.1]

Clark, C. W. and M. Mangel. (1984). Foraging and flocking strategies: information in an uncertain environment. *Am. Nat.* 123:626–641. [4.1]

Cook, R. M. and B. J. Cockrell. (1978). Predator ingestion rate and its bearing on feeding time and the theory of optimal diets. *J. Anim. Ecol.* 46:115–125. [Preface, 2.1, 3.3, Table 9.1]

Coombs, C. H. (1969). Portfolio theory: a theory of risky decision making. In: *La Decision: agrégation et dynamique des order de preference.* Editions du CNRS, Paris. [6.4]

Corbet, S. A., I. Cuthill, M. Fallows, T. Harrison, and G. Hartley. (1981). Why do nectar-foraging bees and wasps work upwards on inflorescences? *Oecologia* 51:79–83. [Table 9.1]

Covich, A. (1972). Ecological economics of seed consumption by Peromyscus: a graphical model of resource substitution. *Trans. Conn. Acad. Arts and Sci.* 44:71–93. [5.1, 5.5]

Cowie, R. J. (1977). Optimal foraging in great tits (*Parus major*). *Nature* 268:137–139. [2.3, 9.2, 9.4, Table 9.1]

Cowie, R. J. and J. R. Krebs. (1979). Optimal foraging in patchy environments. In: R. M. Anderson, B. D. Turner, and L. R. Taylor (eds.), *The British Ecological Society Symposium*, Vol. 20. *Population Dynamics*, pp. 183–205. Blackwell Scientific Publications, Oxford. [8.3]

Cox, D. R. (1962). *Renewal Theory.* Metheuen and Co., London. [Box 3.3]

Crawley, M. J. (1983). *Herbivory.* Blackwell Scientific Publications, Oxford. [5.5]

Curio, E. B. (1973). Towards a methodology of teleonomy. *Experientia* 29:1045–1058. [10.2]

Cuthill, I. C. (1985). Experimental studies of optimal foraging theory. D.Phil. thesis, Oxford University. [9.5]

Darwin, C. R. (1859). *On the Origin of Species*. John Murray, London. [Preface].

Davidson, D. W. (1978). Experimental tests of optimal diet in two social insects. *Behav. Ecol. Sociobiol.* 4:35–41. [Table 9.1].

Davies, N. B. (1977a). Prey selection and social behaviour in wagtails (Aves: Motacillidae). *J. Anim. Ecol.* 46:37–57. [Table 9.1].

Davies, N. B. (1977b). Prey selection and the search strategy of the spotted flycatcher (*Muscicapa striata*): a field study of optimal foraging. *Anim. Behav.* 25: 1016–1033. [Table 9.1]

Davies, N. B. and A. I. Houston. (1981). Owners and satellites: the economics of territory defence in the pied wagtail, *Motacilla alba*. *J. Anim. Ecol.* 50:157–180 [5.4, 8.5]

Davison, M. (1969). Preference for mixed-interval vs. fixed-interval schedules. *J. Exp. Anal. Behav.* 12:247–252. [Table 6.1]

Dawkins, M. S. (1984). Battery hens name their price: consumer demand theory and the measurement of ethological "needs." *Anim. Behav.* 31:1195–1205. [5.1]

Dawkins, R. (1982). *The Extended Phenotype*. W. H. Freeman and Co., Oxford. [8.1]

Dawkins, R. (1983). Adaptationism was always predictive and needed no defense. *Behav. Brain Sci.* 6:360–361. [10.2]

DeBenedictis, P. A., F. B. Gill, F. R. Hainsworth, G. H. Pyke, and L. L. Wolf. (1978). Optimal meal size in hummingbirds. *Am. Nat.* 112:301–316. [2.6]

DeGroot, M. H. (1970). *Optimal Statistical Decisions*. McGraw-Hill, New York. [4.2, 4.4, Box 4.1, 6.2]

Dennett, D. C. (1983). Intentional systems in cognitive ethology: the Panglossian paradigm defended. *Behav. Brain Sci.* 6:343–390. [8.2]

Denno, R. F. (1983). Tracking variable host plants in space and time. In: R. F. Denno and M. S. McClure (eds.), *Variable Plants and Herbivores in Natural and Managed Systems*, pp. 291–341. Academic Press, New York. [5.5]

Denno, R. F. and M. S. McClure (eds.). (1983). *Variable Plants and Herbivores in Natural and Managed Systems*. Academic Press, New York. [5.5]

Dixit, A. K. (1976). *Optimization in Economics*. Oxford University Press, Oxford. [7.2, Box 7.2]

Draulans, D. (1982). Foraging and size selection of mussels by the tufted duck *Aythya fuligula*. *J. Anim. Ecol.* 51:943–956. [Table 9.1]

Ebersole, J. P. and J. C. Wilson. (1980). Optimal foraging: the response of *Peromyscus leucopus* to experimental changes in processing time and hunger. *Oecologia* 46:80–85. [9.5, Table 9.1]

Egan, J. P. (1975). *Signal Detection Theory and ROC Analysis*. Academic Press, New York. [3.7]

Elner, R. W. and R. N. Hughes. (1978). Energy maximization in the diet of the shore crab, *Carcinus maenus*. *J. Anim. Ecol.* 47:103–116. [3.7, Table 3.1, 4.2, 4.5, 9.3, Table 9.1]

Emlen, J. M. (1966). The role of time and energy in food preference. *Am. Nat.* 100: 611–617 [9.1]

Emlen, J. M. (1973). *Ecology: An Evolutionary Approach.* Addison-Wesley, New York. [2.2]

Emlen, J. M. and M. G. R. Emlen. (1975). Optimal choice in diet: test of a hypothesis. *Am. Nat.* 109:427–435. [Table 9.1]

Engen, S. and N. C. Stenseth. (1984). A general version of optimal foraging theory: the effect of simultaneous encounters. *Theor. Popul. Biol.* 26:192–204. [3.2, Box 3.1, Table 3.1]

Erichsen, J. T., J. R. Krebs, and A. I. Houston. (1980). Optimal foraging and cryptic prey. *J. Anim. Ecol.* 49:271–276. [3.7, Table 3.1, 4.2, Table 9.1]

Erlinge, S. (1981). Food preference, optimal diet and reproductive output in stoats (*Mustela erminea*) in Sweden. *Oikos* 36:303–315. [Table 9.1]

Estabrook, G. F. and D. C. Jespersen. (1974). The strategy for a predator encountering a model-mimic system. *Am. Nat.* 108:443–457. [4.1, 4.3]

Evans, R. M. (1982). Efficient use of food patches at different distances from a breeding colony in black-billed gulls. *Behaviour* 79:28–38. [2.6]

Feeney, P. (1975). Biochemical coevolution between plants and their insect herbivores. In: L. E. Gilbert and P. H. Raven (eds.), *Coevolution of Plants and Animals*, pp. 1–19. University of Texas Press, Austin, TX. [5.5]

Finney, D. J. (1962). *Probit Analysis.* Cambridge University Press, New York. [Box 2.3]

Formanowicz, D. R., Jr. (1984). Foraging tactics of an aquatic insect: partial consumption of prey. *Anim. Behav.* 32:774–781. [9.4, Table 9.1]

Freeland, W. J. and D. H. Janzen. (1974). Strategies in herbivory by mammals: the role of plant secondary compounds. *Am. Nat.* 108:269–289. [5.5]

Fretwell. S. D. (1972). *Populations in a Seasonal Environment.* Princeton University Press, Princeton, NJ. [2.6, 9.1]

Friedman, M. and L. J. Savage. (1948). The utility analysis of choices involving risk. *J. Political Economy* 56:279–304. [6.2]

Furnass, T. I. (1979). Laboratory experiments on prey selection by perch fry (*Perca fluviatilis*). *Freshwater Biol.* 9:33–43. [Table 9.1]

Gardner, M. B. (1981). Mechanisms of size selectivity by planktivorous fish: a test of hypotheses. *Ecology* 62:571–578. [Table 9.1]

Georgian, T. and J. B. Wallace. (1981). A model of seston capture by net-spinning caddisflies. *Oikos* 36:147–157. [1.1]

Getty, T. (1985). Discriminability and the sigmoid functional response: how optimal foragers could stabilize model-mimic complexes. *Am. Nat.* 125:239–256. [3.7, Table 3.1]

Getty, T. and J. R. Krebs. (1985). Lagging partial preferences for cryptic prey: a signal detection analysis of great tit foraging. *Am. Nat.* 125:39–60. [3.7, Table 3.1, 4.2, Table 9.1]

Ghiselin, M. (1983). Lloyd Morgan's canon in evolutionary context. *Behav. Brain Sci.* 6:362–363. [9.3, 10.1]

Gibb, J. A. (1958). Predation by tits and squirrels on the eucosmid *Ernarmonia conicolana* (Heyl.). *J. Anim. Ecol.* 27:375–396. [8.3]

Gibson, R. M. (1980). Optimal prey-size selection by three-spined sticklebacks (*Gasterosteus aculeatus*): a test of the apparent-size hypothesis. *Z. Tierpsychol.* 52:291–307. [Table 9.1]

Gill, F. B. and L. L. Wolf. (1975). Economics of feeding territoriality in the golden-winged sunbird. *Ecology* 56:333–345. [9.5]

Giller, P. S. (1980). The control of handling time and its effects on the foraging strategy of a heteropteran predator, Notonecta. *J. Anim. Ecol.* 49:699–712. [3.3, Table 9.1]

Gilliam, J. F. (1982). Foraging under mortality risk in size-structured populations. Ph.D. diss., Michigan State University. [7.3]

Gilliam J. F., R. F. Green, and N. E. Pearson. (1982). The fallacy of the traffic policeman: a response to Templeton and Lawlor. *Am. Nat.* 119:875–878. [2.1]

Giraldeau, L-A. and D. L. Kramer. (1982). The marginal value theorem: a quantitative test using load size variation in the central place forager, the eastern chipmunk, *Tamias striatus. Anim. Behav.* 30:1036–1042. [2.3, Table 9.1]

Gittelmann, S. H. (1978). Optimum diet and body size in backswimmers (Heteroptera: Notonectidae, Pleidae). *Ann. Ent. Soc. Am.* 71:737–747. [Table 9.1]

Glander, K. E. (1981). Feeding patterns in mantled howler monkeys. In: A. C. Kamil and T. D. Sargent (eds.), *Foraging Behavior: Ecological, Ethological and Psychological approaches*, pp. 231–257. Garland STPM Press, New York. [5.5]

Godin, J-G. J. and M. H. A. Keenleyside. (1984). Foraging on patchily distributed prey by a cichlid fish (Telostei, Cichlidae): a test of the ideal free distribution theory. *Anim. Behav.* 32:120–131. [9.1]

Goss-Custard, J. D. (1977a). Optimal foraging and the size selection of worms by redshank, *Tringia totanus*, in the field. *Anim. Behav.* 25:10–29. [3.2, Table 9.1]

Goss-Custard, J. D. (1977b). Predator responses and prey mortality in the redshank, *Tringia totanus* (L.), and a preferred prey, *Corophium volutator* (Pallus). *J. Anim. Ecol.* 46:21–35. [Table 9.1]

Gould, J. P. (1974). Risk, stochastic preference, and the value of information. *J. Econ. Theory* 8:64–84. [4.2]

Gould, S. J. (1978). *Ever Since Darwin*. Andre Deutsch, London. [10.1]

Gould, S. J. (1980). *The Panda's Thumb*. W. W. Norton, New York. [8.1, 10.1]

Gould, S. J. (1981). *The Mismeasure of Man*. W. W. Norton, New York. [10.1]

Gould, S. J. and R. C. Lewontin. (1979). The spandrels of San Marco and the Panglossian paradigm: a critique of the adaptationist programme. *Proc. R. Soc. London* (B) 205:581–598. [Preface, 8.1, 10.1, 10.2]

Gould, S. J. and E. S. Vrba. (1982). Exaptation—a missing term in the science of form. *Paleobiology* 8:4–15. [10.2]

Gray, R. (1986). Faith and foraging. In: A. C. Kamil, J. R. Krebs, and H. R. Pulliam (eds.), *Foraging Behavior*. Plenum Press, New York [9.3]

Green, L., E. B. Fisher Jr., S. Perlow, and L. Sherman. (1981). Preference reversal and self-control: choice as a function of reward amount and delay. *Behav. Anal. Letters* 1:244–256. [Box 6.2]

Green, R. F. (1980). Bayesian birds: a simple example of Oaten's stochastic model of optimal foraging. *Theor. Popul. Biol.* 18:244–256. [2.3, 2.4, Box 2.1, 4.1, 4.4, 9.5]

Green, R. F. (1984). Stopping rules for optimal foragers. *Am. Nat.* 123:30–40. [4.1, 4.4, 8.3]

Griffiths, D. (1982). Tests of alternative models of prey consumption by predators, using ant-lion larvae. *J. Anim. Ecol.* 52:363–373. [Table 9.1]

Hairston, N. G., F. E. Smith, and L. B. Slobodkin. (1960). Community structure, population control and competition. *Am. Nat.* 94:421–424. [5.5]

Hames, R. B. and W. T. Vickers. (1982). Optimal diet breadth theory as a model to explain variability in Amazonian hunting. *Amer. Ethnol.* 9:258–278. [Table 9.1]

Harborne, J. B. (1978). *Biochemical Aspects of Plant and Animal Coevolution.* Academic Press, London. [5.5]

Harborne, J. B. (1982). *Introduction to Ecological Biochemistry*, 2d ed. Academic Press, London. [5.5]

Harley, C. B. (1981). Learning the evolutionarily stable strategy. *J. Theor. Biol.* 89:611–633. [4.1]

Harper, D. G. C. (1982). Competitive foraging in mallards: ideal free ducks. *Anim. Behav.* 30:575–584. [9.1]

Hartling, L. K. and R. C. Plowright. (1979). Foraging by bumblebees on patches of artificial flowers: a laboratory study. *Can. J. Zool.* 57:1866–1870. [Table 9.1]

Hartwick, E. B. (1976). Foraging strategy of the black oystercatcher (*Haematopus bachmani* Audubon). *Can. J. Zool.* 54:142–155. [Table 9.1]

Hassell, M. P. (1980). Foraging strategies, population models and biological control: a case study. *J. Anim. Ecol.* 49:603–628. [Table 9.1]

Hassell, M. P., J. H. Lawton, and J. R. Beddington. (1976). Components of arthropod I: the prey death rate. *J. Anim. Ecol.* 45:135–164. [3.3]

Hawkes, K., K. Hill, and J. O'Connell. (1982). Why hunters gather: optimal foraging and the Ache of eastern Paraguay. *Amer. Ethnol.* 9:379–398. [Table 9.1]

Hawkes, K., J. O'Connell, K. Hill, and E. Charnov. (1985). How much is enough? Hunters and limited needs. *Ethology and Sociobiol.* 6:23–35. [Box 5.1]

Haynes, J. and M. Mesler. (1984). Pollen foraging by bumblebees: foraging patterns and efficiency of *Lupinus polyphyllus*. *Oecologia* 63:357–363. [Table 9.1]

Heads, P. A. and J. H. Lawton. (1983). Studies on the natural enemy complex of the holly leaf miner: the effects of scale on the detection of aggregative responses and the implications for biological control. *Oikos* 40:267–276. [Table 9.1]

Hegner, R. E. (1982). Central place foraging in the white-fronted bee-eater. *Anim. Behav.* 30:953–963. [Table 9.1]

Heinrich, B. (1979). Resource heterogeneity and patterns of movement in foraging bumblebees. *Oecologia* 40:235–245. [5.5]

Heller, R. (1980). On optimal diet in a patchy environment. *Theor. Popul. Biol.* 17:201–214. [3.2]

Heller, R. and M. Milinski. (1979). Optimal foraging of sticklebacks on swarming prey. *Anim. Behav.* 27:1127–1141. [7.3]

Henderson, J. M. and R. E. Quandt. (1971). *Microeconomic Theory: A Mathematical Approach.* McGraw-Hill, New York. [5.2]

Herrnstein, R. J. (1964). Aperiodicity as a factor in choice. *J. Exp. Anal. Behav.* 7:179–182. [Table 6.1]

Herrnstein, R. J. (1970). On the law of effect. *J. Exp. Anal. Behav.* 13:243–266. [8.5]

Herrnstein, R. J. and D. H. Loveland. (1975). Maximizing and matching on concurrent ratio schedules. *J. Exp. Anal. Behav.* 24:107–116. [5.3]

Herrnstein, R. J. and W. Vaughan Jr. (1980). Melioration and behavioral allocation. In: J. E. R. Staddon (ed.), *Limits to Action: The Allocation of Individual Behavior*, pp. 143–176. Academic Press, New York. [8.5]

Hey, J. D. (1979). *Uncertainty in Microeconomics.* Martin Robertson, Oxford. [6.2]

Heyman, G. M. and R. D. Luce. (1979). Operant matching is not a logical consequence of maximizing reinforcement rate. *Anim. Learning and Behav.* 7:133–140. [8.5]

Hinde, R. A. (1975). The concept of function. In: G. Baerends, C. Beer, and A. Manning (eds.), *Function and Evolution in Behaviour.* Clarendon Press, Oxford. pp. 3–15. [10.2]

Hinson, J. M. and J. E. R. Staddon. (1983). Hill-climbing by pigeons. *J. Exp. Anal. Behav.* 39:25–47. [8.5]

Hirshleifer, J. (1966). Investment decision under uncertainty: application of the state-preference approach. *Quart. J. Econ.* 80:252–277. [6.4]

Hodges, C. M. (1981). Optimal foraging in bumblebees: hunting by expectation. *Anim. Behav.* 29:1166–1171. [9.5, Table 9.1]

Hodges, C. M. (1985). Bumblebee foraging: the threshold departure rule. *Ecology* 66:179–187. [Table 9.1]

Hodges, C. M. and L. L. Wolf. (1981). Optimal foraging bumblebees: why is nectar left behind in flowers? *Behav. Ecol. Sociobiol.* 9:41–44. [Table 9.1]

Hogan, J. A., S. Kleist, and C. S. L. Hutchings. (1970). Display and food as reinforcers in the Siamese fighting fish (*Betta splendens*). *J. Comp. Physiol. Psychol.* 70:351–357. [5.2]

Holling, C. S. (1959). Some characteristics of simple types of predation and parasitism. *Can. Entomol.* 91:385–398. [2.1]

Holling, C. S. (1966). The strategy of building models of complex ecological systems. In: K. E. F. Watt (ed.), *Systems Analysis in Ecology*, pp. 195—214. Academic Press, New York. [9.1]

Horn, H. S. (1979). Adaptation from the perspective of optimality. In: O. T. Solbrig, S. Jain, G. B. Johnson, and P. H. Raven (eds.), *Topics in Plant Population Biology*, pp. 48–61. Columbia University Press, New York. [10.4]

Horn, M. H. (1983). Optimal diets in complex environments—feeding strategies of two herbivorous fishes from a temperate rocky intertidal zone. *Oecologia* 58:345–350. [Table 9.1]

Houston, A. I. (1983). Optimality and matching. *Behav. Anal. Letters* 3:1–15. [8.5]

Houston, A. I. (1986). Control of foraging decisions. In: M. L. Commons, S. J. Shettleworth, and A. Kacelnik (eds.), *Quantitative Analyses of Behavior*, Vol. 6. *Foraging*. Erlbaum, New York. [Preface]

Houston, A. I., A. Kacelnik, and J. M. McNamara. (1982). Some learning rules for acquiring information. In: D. J. McFarland (ed.), *Functional Ontogeny*, pp. 140–191. Pitman Books, London. [1.4, 4.1]

Houston, A. I., J. R. Krebs, and J. T. Erichsen. (1980). Optimal prey choice and discrimination time in the great tit (*Parus major* L.). *Behav. Ecol. Sociobiol.* 6:169–175. [3.7, Table 3.1, 4.2, Table 9.1]

Houston, A. I. and J. M. McNamara. (1981). How to maximize reward rate on two variable-interval paradigms. *J. Exp. Anal. Behav.* 35:367–396. [8.5]

Houston, A. I. and J. M. McNamara. (1982). A sequential approach to risk-taking. *Anim. Behav.* 30:1260–1261. [6.4, 7.1]

Houston, A. I. and J. M. McNamara. (1985). The variability of behaviour and constrained optimization. *J. Theor. Biol.* 112:265–273. [9.1]

Hubbard, S. F. and R. M. Cook. (1978). Optimal foraging by parasitoid wasps. *J. Anim. Ecol.* 47:593–604. [Table 9.1]

Hughes, R. N. (1979). Optimal diets under the energy maximization premise: the effects of recognition time and learning. *Am. Nat.* 113:209–221. [3.7]

Hughes, R. N. and R. W. Elner. (1979). Tactics of a predator, *Carcinus maenus*, and morphological responses of the prey, *Nucella lapillus*. *J. Anim. Ecol.* 48:65–78. [Table 9.1]

Hughes, R. N. and R. Seed. (1981). Size selection of mussels by the blue crab *Callinectes:* energy maximizer? *Mar. Ecol. Prog. Ser.* 6:83–89. [Table 9.1]

Iwasa, Y., M. Higashi, and N. Yamamura. (1981). Prey distribution as a factor determining the choice of optimal foraging strategy. *Am. Nat.* 117:710–723. [4.1, 8.3]

Jacobs, O. L. R. (1974). *Introduction to Control Theory*. Clarendon Press, Oxford. [7.2, 7.4, Box 7.1]

Jaeger, R. G. and D. E. Barnard. (1981). Foraging tactics of a terrestrial salamander: choice of diet in structurally simple environments. *Am. Nat.* 117:639–664. [Table 9.1]

Jaeger, R. G. and A. M. Rubin. (1981). Foraging tactics of a terrestrial salamander— judging prey profitability. *J. Anim. Ecol.* 51:167–176. [Table 9.1]

Jaenike, J. (1978). On optimal oviposition behavior in phytophagous insects. *Theor. Popul. Biol.* 14:350–356. [5.5]

Janetos, A. C. and B. J. Cole. (1981). Imperfectly optimal animals. *Behav. Ecol. Sociobiol.* 9:203–210. [1.5, 8.2]

Jenkins, S. H. (1980). A size-distance relation in food selection by beavers. *Ecology* 61:740–746. [Table 9.1]

Kacelnik, A. (1984). Central place foraging in starlings (*Sturnus vulgaris*). 1. Patch residence time. *J. Anim. Ecol.* 53:283–300. [9.2, 9.4, Table 9.1]

Kacelnik, A. and I. Cuthill. (1986). Optimal foraging: just a matter of technique. In: A. C. Kamil, J. R. Krebs, and H. R. Pulliam (eds.), *Foraging Behavior*. Plenum Press, New York. [1.2, 4.4, 9.5]

Kacelnik A. and A. I. Houston. (1984). Some effects of energy costs on foraging strategies. *Anim. Behav.* 32:609–614. [2.3, 9.4, Table 9.1]

Kacelnik, A., A. I. Houston, and J. R. Krebs. (1981). Optimal foraging and territorial defence in the great tit (*Parus major*). *Behav. Ecol. Sociobiol.* 8:35–40. [5.3]

Kacelnik A. and J. R. Krebs. (1983). The dawn chorus in the great tit (*Parus major*): proximate and ultimate causes. *Behaviour* 82:287–309. [7.3]

Kacelnik, A. and J. R. Krebs. (1985). Learning to exploit patchily distributed food. In: R. M. Sibly and R. H. Smith (eds.), *Behavioural Ecology, 25th Symposium of the British Ecological Society*, pp. 189–205. Blackwell Scientific Publications, Oxford. [4.1, 4.3]

Kagel, J. H., L. Green, and T. Caraco. (1986). When foragers discount the future: constraint or adaptation? *Anim. Behav.* 34:271–283. [6.6, Box 6.2]

Kamil, A. C. (1978). Systematic foraging for nectar by amakihi, *Loxops virens*. *J. Comp. Physiol. Psychol.* 92:388–396. [8.5]

Kamil, A. C. (1983). Optimal foraging theory and the psychology of learning. *Am. Zool.* 23:291–302. [8.5]

Kamil, A. C. and S. J. Yoerg. (1982). Learning and optimal foraging. In: P. P. G. Bateson and P. H. Klopfer (eds.), *Perspectives in Ethology*, Vol. 5, pp. 325–364. Plenum, New York. [8.5]

Kasuya, E. (1982). Central place water collection in a Japanese paper wasp (*Polistes chinensis antennalis*). *Anim. Behav.* 30:1010–1014. [Table 9.1]

Kaufman, L. W. and G. Collier. (1981). Economics of seed handling. *Am. Nat.* 118:46–60. [Table 9.1]

Killeen, P. R. (1981). Averaging theory. In: C. M. Bradshaw, E. Szabadi, and C. F. Lowe (eds.), *Quantification of Steady State Operant Behaviour*, pp. 21–34. Elsevier/North Holland, Amsterdam. [4.1]

Killeen, P. R., J. P. Smith, and S. J. Hanson. (1981). Central place foraging in *Rattus norvegicus*. *Anim. Behav.* 29:64–70. [Table 9.1]

Kislalioglu, M. and R. N. Gibson. (1976). Some factors governing prey selection by the 15-spined stickleback, *Spinachia spinachia* (L.). *J. Exp. Mar. Biol. Ecol.* 25:159–169. [Table 9.1]

Krebs, J. R. (1973a). Behavioral aspects of predation. In: P. P. G. Bateson and P. H. Klopfer (eds.), *Perspectives in Ethology*, Vol. 1, pp. 73–111. Plenum, New York. [8.3]

Krebs, J. R. (1973b). Analysis of function. Review of *Function and evolution in behaviour*, edited by G. P. Baerends, C. Beer, and A. Manning. Clarendon Press, Oxford. *Nature* 260:196–197. [10.2]

Krebs, J. R. and M. I. Avery. (1984). Diet and nestling growth in the European bee-eater. *Oecologia* 64:363–368. [5.5]

Krebs, J. R. and M. I. Avery. (1985). Test of central place foraging in a single-prey loader. *J. Anim. Ecol.* 54:459–472. [3.5, Table 9.1]

Krebs, J. R., J. T. Erichsen, M. I. Webber, and E. L. Charnov. (1977). Optimal prey-selection by the great tit (*Parus major*). *Anim. Behav.* 25:30–38. [Box 2.3, 3.2, 8.6, 9.4, Table 9.1]

Krebs, J. R., A. Kacelnik, and P. Taylor. (1978). Test of optimal sampling by foraging great tits. *Nature* 275:27–31. [4.1, 4.3].

Krebs, J. R. and R. H. McCleery. (1984). Optimization in behavioural ecology. In: J. R. Krebs and N. B. Davies (eds.), *Behavioural Ecology: An Evolutionary Approach*, 2d ed., pp. 91–121. Sinauer Associates, Sunderland, MA. [Box 2.3, 3.2, 8.6, 9.5]

Krebs, J. R., J. C. Ryan, and E. L. Charnov. (1974). Hunting by expectation or optimal foraging? A study of patch use by chickadees. *Anim. Behav.* 22:953–964. [8.3, Table 9.1]

Krebs, J. R., D. W. Stephens, and W. J. Sutherland. (1983). Perspectives in optimal foraging. In: A. H. Brush and G. A. Clark, Jr. (eds.), pp. 165–216. Cambridge University Press, New York. [8.2, 9.3, 9.5]

Lacher, T. E., M. R. Willing, and M. A. Mares. (1982). Food preference as a function of resource abundance with multiple prey types: an experimental analysis of optimal foraging theory. *Am. Nat.* 120:297–316. [Table 9.1]

Larkin, S. B. C. (1981). Time and energy in decision making. D. Phil. thesis, Oxford University. [Table 9.1]

Lea, S. E. G. (1978). The psychology and economics of demand. *Psychol. Bull.* 85:441–466. [5.1, 5.3]

Lea, S. E. G. (1979). Foraging and reinforcement schedules in the pigeon: optimal and non-optimal aspects of choice. *Anim. Behav.* 27:875–886. [Table 9.1]

Lea, S. E. G. (1982). The mechanism of optimality in foraging. In: M. L. Commons, R. J. Herrnstein, and H. Rachlin (eds.), *Quantitative Analyses of Behavior, 2: Matching and Maximizing Accounts*, pp. 355–406. Ballinger, Cambridge, MA. [8.5]

Lea, S. E. G. (1983). The analysis of need. In: R. L. Mellgren (ed.), *Animal Cognition and Behavior*, pp. 31–63. North Holland, New York. [5.1, 5.3]

Lessells, C. M. and D. W. Stephens. (1983). Central place foraging: single-prey loaders again. *Anim. Behav.* 31:238–243. [3.5, Box 3.4, Table 3.1]

Lester, N. P. (1984). The feed-feed decision: how goldfish solve the patch depletion problem. *Behaviour* 85:175–199. [4.1, 9.3]

Levander, O. A. and V. C. Morris. (1970). Interactions of methione, vitamin E, and antioxidants in selenium toxicity in the rat. *J. Nutrition* 100:1111–1118. [5.5]

Leventhal, A. M., R. F. Morrell, E. F. Morgan Jr., and C. C. Perkins, Jr. (1959). The relation between mean reward and mean reinforcement. *J. Exp. Psychol.* 57:284–287. [Table 6.1]

Lewis, A. R. (1980). Patch use by gray squirrels and optimal foraging. *Ecology* 61:1371–1379. [Table 9.1]

Lewis, A. R. (1982). Selection of nuts by gray squirrels and optimal foraging theory. *Amer. Midl. Nat.* 107:250–257. [Table 9.1]

Lewontin, R. C. (1978). Adaptation. *Scient. Am.* 239:156–169. [10.1, 10.2]

Lewontin, R. C. (1979). Fitness, survival, and optimality. In: D. J. Horn, G. R. Stairs, and R. D. Mitchell (eds.), *The Analysis of Ecological Systems*, pp. 3–21. Ohio State University Press, Columbus. [Preface, 10.2]

Lewontin, R. C. (1983). Elementary errors about evolution. *Behav. Brain Sci.* 6:367–368. [10.1]

Lewontin, R. C. (1984). Adaptation. In: E. Sober (ed.), *Conceptual Issues in Evolutionary Biology*, pp. 237–251. MIT Press, Cambridge, MA. (Reprinted from *Encyclopedia Einaudi*, Torino, 1977.) [Preface, 10.2]

Lima, S. L. (1983). Downy woodpecker foraging behavior: efficient sampling in simple stochastic environments. *Ecology* 65:166–174. [4.1, 4.4]

Lima, S. L. (1985a). Sampling behavior of starlings foraging in simple patchy environments. *Behav. Ecol. Sociobiol.* 16:135–142. [4.1, 4.4]

Lima, S. L. (1985b). Maximizing feeding efficiency and minimizing time exposed to predators: a tradeoff in the black-capped chickadee. *Oecologia* 66:60–67. [Box 5.2]

Lima, S. L., T. J. Valone, and T. Caraco. (1985). Foraging efficiency-predation risk trade-off in the grey squirrel. *Anim. Behav.* 33:155–165. [Box 5.2]

Lind, H. (1965). Parental feeding in the oystercatcher (*Haematopus o. ostralegus* L.). *Dansk. Orn. Foren. Tidsskr.* 59:1–31. [Table 9.1]

Lobel, P. S. and J. C. Ogden. (1981). Foraging by the herbivorous parrot fish *Sparisoma radians. Mar. Biol.* 64:173–183. [Table 9.1]

Lorenz, K. (1949). Die Beziehungen zwischen Kopform und Zirkelbewegung bie Sturndien und Icteridien. In: E. Mayr and E. Schuz (eds.), *Ornithologe als Biologische Wissenschaft*, pp. 153–157. Carl Winter, Heidelberg. [3.5]

Lucas, J. R. (1983). The role of foraging time constraints and variable prey encounter in optimal diet choice. *Am. Nat.* 122:191–209. [3.3, 3.4, Table 3.1]

Lucas, J. R. (1985). Partial prey consumption by antlion larvae. *Anim. Behav.* 33:945–959. [2.1]

Lucas, J. R. and A. Grafen. (1985). Partial prey consumption by ambush predators. *J. Theor. Biol.* 113:455–473. [3.3, Table 3.1]

Luria, S. E., S. J. Gould, and S. Singer. (1981). *A View of Life.* Benjamin/Cummings, Menlo Park, CA. [10.2]

MacArthur, R. H. and E. R. Pianka. (1966). On optimal use of a patchy environment. *Am. Nat.* 100:603–609. [2.1, 9.1]

Macevicz, S. and G. Oster. (1976). Modelling social insect populations II. Optimal reproductive strategies in annual eusocial insect colonies. *Behav. Ecol. Sociobiol.* 1:265–282. [7.3]

Maderson, P. F. A. (1982). The role of development in macroevolutionary change. In: J. T. Bonner (ed.), *Evolution and Development.* Dahlem Conference Report No. 22, pp. 279–312. Springer-Verlag, Berlin. [8.1]

Mangel, M. and C. W. Clark. (1986). Unified foraging theory. *Ecology.* [7.4]

Mansfield, E. (1979). *Microeconomics: Theory and Applications.* W. W. Norton. New York. [5.2]

Marden, J. H. and K. D. Waddington. (1981). Floral choices by honeybees in relation to the relative distances of flowers. *Physiol. Entomol.* 6:431–435. [Table 9.1]

Martin, J. (1983). Optimal foraging theory: a review of some models and their applications. *Am. Anthropologist* 85:612–629. [3.7]

Maynard Smith, J. (1972). *On Evolution.* Edinburgh University Press, Edinburgh. [10.2]

Maynard Smith, J. (1974). *Models in Ecology.* Cambridge University Press, Cambridge. [2.2]

Maynard Smith, J. (1978). Optimization theory in evolution. *Ann. Rev. Ecol. Syst.* 9:31–56. [1.4, 5.3]

Maynard Smith, J. (1982). *Evolution and the Theory of Games.* Cambridge University Press, Cambridge. [1.4, 10.2]

Mayr, E. (1982). Adaptation and selection. *Biol. Zentralbl.* 101:66–77. [8.1, 10.2]

Mayr, E. (1983). How to carry out the adaptationist program? *Am. Nat.* 121: 324–334. [8.1, 10.2]

Mazur, J. E. (1981). Optimization theory fails to predict performance of pigeons in two-response situation. *Science* 214:823–825. [8.5]

McFarland, D. J. (1977). Decision-making in animals. *Nature* 269:15–21. [9.1]

McFarland, D. J. and A. I. Houston. (1981). *Quantitative Ethology: The State Space Approach.* Pitman, London. [1.4, 5.1, 5.2, Box 6.2, 7.4, Box 7.1, 10.1]

McGinley, M. A. (1984). Central place foraging for non-food items: determination of the stick size-value relationship of housebuilding materials collected by eastern wood rats. *Am. Nat.* 123:841–853. [Table 9.1]

McKey, D. (1979). The distribution of secondary compounds within plants. In: G. A. Rosenthal and D. H. Janzen (eds.), *Herbivores,* pp. 56–133. Academic Press, New York. [5.5]

McNair, J. N. (1979). A generalized model of optimal diets. *Theor. Popul. Biol.* 15:159–170. [2.6, Box 2.1, 3.4, Box 3.3, Table 3.1]

McNair, J. N. (1982). Optimal giving-up times and the marginal value theorem. *Am. Nat.* 119:511–529. [8.3, Table 9.1]

McNair, J. N. (1983). A class of patch-use strategies. *Am. Zool.* 23:303–313. [3.3, Box 3.2, Table 3.1]

McNamara, J. M. (1982). Optimal patch use in a stochastic environment. *Theor. Popul. Biol.* 21:269–288. [2.3, 4.1, 4.4, 8.3]

McNamara, J. M. (1983). Optimal control of the diffusion coefficient of a simple diffusion process. *Math. Oper. Res.* 8:373–380. [6.4, 7.1]

McNamara, J. M. (1984). Control of a diffusion by switching between two drift-diffusion coefficient pairs. *SIAM J. Control and Optimization* 22:87–94. [6.4, 7.1]

McNamara, J. M. and A. I. Houston. (1980). The application of statistical decision theory to animal behavior. *J. Theor. Biol.* 85:673–690. [4.1, 4.3, Box 4.1]

McNamara, J. M. and A. I. Houston. (1982). Short-term behaviour and lifetime fitness. In: D. J. McFarland (ed.), *Functional Ontogeny,* pp. 60–87. Pitman, London. [6.4]

McNeill, S. and T. R. E. Southwood. (1978). The role of nitrogen in the development of insect/plant relationships. In: J. B. Harborne (ed.), *Biochemical Aspects of Plant and Animal Coevolution,* pp. 77–98. Academic Press, London. [5.5]

Mellgren, R. L. (1982). Foraging in a simulated natural environment: there's a rat loose in the lab. *J. Exp. Anal. Behav.* 38:93–100. [Table 9.1]

Mellgren, R. L., L. Misasi, and S. W. Brown. (1984). Optimal foraging theory: prey density and travel requirements in *Rattus norvegicus. J. Comp. Psychol.* 98:142–153. [Table 9.1]

Milinski, M. (1979). An evolutionarily stable feeding strategy in sticklebacks. *Z. Tierpsychol.* 51:36–40. [9.1]

Milinski, M. and R. Heller. (1978). Influence of a predator on the optimal foraging behaviour of sticklebacks (*Gasterosteus aculeatus L.*). *Nature* 275:642–644. [7.3]

Mills, S. and J. Beatty. (1979). The propensity interpretation of fitness. *Phil. Sci.* 46:263–286. [Preface]

Milton, K. (1979). Factors influencing leaf choice by howler monkeys: a test of some hypotheses of food selection by generalist herbivores. *Am. Nat.* 114:362–378. [Table 9.1]

Mitchell, R. (1981). Insect behavior, resource exploitation, and fitness. *Ann. Rev. Ent.* 26:373–396. [5.5]

Mittelbach, G. G. (1981). Foraging efficiency and bodysize: a study of optimal diet and habitat used by bluegills. *Ecology* 62:1370–1386. [Table 9.1]

Moermond, T. and J. Denslow. (1984). Fruit choice in Neotropical birds. *J. Anim. Ecol.* 52:407–420. [Table 9.1]

Montgomerie, R. D., J. McA. Eadie, and L. D. Harder. (1984). What do foraging hummingbirds maximize? *Oecologia* 63:357–363. [Table 9.1]

Morley, F. (ed.). (1981). *Grazing Animals.* Elsevier, Amsterdam. [5.5]

Munger, J. C. (1984). Optimal foraging patch use by horned lizards (Iguanidae: *Phrynosoma*). *Am. Nat.* 123:654–680. [Table 9.1]

Myers, J. P. (1983). Commentary. In: A. H. Brush and G. A. Clark, Jr. (eds.), *Perspectives in Ornithology*, pp. 216–221. Cambridge University Press, New York. [8.2, 8.6, 10.1]

Oaten, A. (1977). Optimal foraging in patches: a case for stochasticity. *Theor. Popul. Biol.* 12:263–285. [2.3, 2.5, Box 2.1, 4.1, 4.4]

O'Brien, W. J., N. A. Slade, and G. L. Vinyard. (1976). Apparent size as the determinant of prey selection by bluegill sunfish (*Lepomis macrochirus*). *Ecology* 57:1204–1310. [3.2]

Ohguchi, O. and K. Aoki. (1983). Effects of colony need for water on optimal food choice in honeybees. *Behav. Ecol. Sociobiol.* 12:77–84. [Table 9.1]

Ollason, J. G. (1980). Learning to forage—optimally? *Theor. Popul. Biol.* 18:44–56. [4.1]

Orians, G. H. (1981). Foraging behavior and the evolution of discriminatory abilities. In: A. C. Kamil and T. D. Sargent (eds.), *Foraging Behavior: Ecological, Ethological and Psychological Approaches*, pp. 389–405. Garland STPM Press, New York [4.1]

Orians, G. H. and N. E. Pearson. (1979). On the theory of central place foraging. In: D. J. Horn, R. D. Mitchell, and G. R. Stairs (eds.), *Analysis of Ecological Systems*, pp. 154–177. Ohio State University Press, Columbus. [3.2, 3.5, Table 3.1, 9.4]

Oster, G. F. and P. Alberch. (1982). Evolution and bifurcation in developmental programs. *Evolution* 36:444–459. [8.1]

Oster, G. F. and E. O. Wilson. (1978). *Caste and Ecology in the Social Insects.* Princeton University Press, Princeton, NJ. [6.5, 10.2]

Owen-Smith, N. and P. Novellie. (1982). What should a clever ungulate eat? *Am. Nat.* 119:151–178. [5.5]

Palmer, A. R. (1984). Prey selection by thadid gastropods: some observational and experimental field tests of foraging models. *Oecologia* 62:162–172. [Table 9.1]

Parker, G. A. (1978). Search for mates. In: J. R. Krebs and N. B. Davies (eds.), *Behavioural Ecology: An Evolutionary Approach*, 1st ed. pp. 214–244. Sinuaer Associates, Sunderland, MA. [Preface, 9.1]

Parker, G. A. and R. A. Stuart. (1976). Animal behavior as a strategy optimizer: evolution of resource assessment strategies and optimal emigration thresholds. *Am. Nat.* 110:1055–1076. [Table 9.1]

Parker, M. A. (1984). Local food depletion and the foraging behavior of a specialist grasshopper *Hesperotettix viridis*. *Ecology* 65:824–835. [5.5, Table 9.1]

Partridge, L. (1981). Increased preferences for familiar foods in small mammals. *Anim. Behav.* 29:211–216. [5.5]

Pastorok, R. A. (1980). The effects of predator hunger and food abundance on prey selection by *Chaoborus* larvae. *Limnol. Oceanogr.* 25:910–921. [Table 9.1]

Pubols, B. H. (1962). Constant versus variable delay of reinforcement. *J. Comp. Physiol. Psychol.* 55:52–56. [Table 6.1]

Pulliam, H. R. (1974). On the theory of optimal diets. *Am. Nat.* 108:59–75. [2.2, Box 2.1]

Pulliam, H. R. (1975). Diet optimization with nutrient constraints. *Am. Nat.* 109: 765–768. [1.5, 3.6, Box 3.5, Table 3.1]

Pulliam, H. R. (1980). Do chipping sparrows forage optimally? *Ardea* 68:75–82. [Box 2.3, Table 9.1]

Pulliam, H. R. (1981). Learning to forage optimally. In: A. C. Kamil and T. D. Sargent (eds.), *Foraging Behavior: Ecological, Ethological and Psychological Approaches*, pp. 379–388. Garland STPM Press, New York. [4.1]

Pulliam, H. R. and C. Dunford. (1980). *Programmed to Learn: An essay on the Evolution of Culture*. Columbia University Press, New York. [4.1]

Pulliam, H. R. and G. C. Millikan. (1982). Social organization in the non-reproductive season. In: D. S. Farner and J. R. King (eds.), *Avian Biology*, Vol. 6, pp. 169–197. Academic Press, New York. [6.4]

Pyke, G. H. (1978a). Optimal foraging in hummingbirds: testing the marginal value theorem. *Am. Zool.* 18:739–752. [Preface, 2.3, 9.5, Table 9.1]

Pyke, G. H. (1978b). Optimal foraging in bumblebees and coevolution with their plants. *Oecologia* 36: 281–293. [Table 9.1]

Pyke, G. H. (1981a). Optimal travel speeds of animals. *Am. Nat.* 118:475–487. [2.6]

Pyke, G. H. (1981b). Optimal foraging in bumblebees: rule of departure from an inflorescence. *Can. J. Zool.* 60:417–428. [4.4]

Pyke, G. H. (1983). Animal movements: an optimal foraging approach. In: I. R. Swingland and P. J. Greenwood (eds.), *The Ecology of Animal Movement*, pp. 7–31. Clarendon Press, Oxford. [2.6]

Pyke, G. H. (1984). Optimal foraging theory: a critical review. *Ann. Rev. Ecol. Syst.* 15:523–575. [9.3]

Pyke, G. H., H. R. Pulliam, and E. L. Charnov. (1977). Optimal foraging: a selective review of theory and tests. *Q. Rev. Biol.* 52:137–154. [1.4, 2.5, 4.1, 4.4]

Rachlin, H., R. Battalio, J. Kagel, and L. Green. (1981). Maximization theory in behavioral psychology. *Behav. Brain Sci.* 4:371–417. [5.1, 5.3]

Rachlin, H., L. Green. J. H. Kagel, and R. C. Battalio. (1976). Economic demand theory and psychological studies of choice. In: G. Bower (ed.), *The Psychology of Learning and Motivation*, Vol. 10, pp. 129–154. Academic Press, New York. [5.3, 5.5]

Raiffa, H. (1968). *Decision Analysis*. Addison–Wesley, Reading, MA. [Box 4.1, 6.2]

Rapport, D. J. (1971). An optimization model of food selection. *Am. Nat.* 105:575–587. [5.1]

Rapport, D. J. (1980). Optimal foraging for complementary resources. *Am. Nat.* 116:324–346. [5.1, 5.5, Table 9.1]

Rapport, D. J. (1981). Foraging behavior of *Stentor coeruleus*: a microeconomic interpretation. In: A. C. Kamil and T. D. Sargent (eds.), *Foraging Behavior: Ecological, Ethological and Psychological Approaches*, pp. 77–83. Garland STPM Press, New York. [5.1, 5.5]

Rapport, D. J. and J. E. Turner. (1977). Economic models in ecology. *Science* 195:367–373. [5.1]

Rausher, M. D. (1983a). Alteration of oviposition behavior by *Battus philenor* butterflies in response to variation in host plant density. *Ecology* 64:1028–1034. [5.5]

Rausher, M. D. (1983b). Conditioning and genetic variation as causes of individual variation in oviposition behavior of the tortoise beetle, *Deloyte guttata*. *Anim. Behav.* 31:743–747. [5.5]

Rausher, M. D. and D. R. Papaj. (1983). Demographic consequences of discrimination among conspecific host plants by *Battus philenor* butterflies. *Ecology* 64:1402-1410. [5.5]

Real, L. A. (1980a). Fitness, uncertainty, and the role of diversification in evolution and behavior. *Am. Nat.* 115:623-638. [6.5, Box 6.1]

Real, L. A. (1980b). On uncertainty and the law of diminishing returns in evolution and behavior. In: J. E. R. Staddon (ed.), *Limits to Action: The Allocation of Individual Behavior*, pp. 37–64. Academic Press, New York. [6.5]

Real, L. A. (1981). Uncertainty and plant-pollinator interactions: the foraging behavior of bees and wasps on artificial flowers. *Ecology* 62:20–26. [Table 6.1]

Real, L. A., J. Ott, and E. Silverfine. (1982). On the tradeoff between mean and variance in foraging: an experimental analysis with bumblebees. *Ecology* 63:1617–1623. [Table 6.1]

Rechten, C., M. I. Avery, and T. A. Stevens. (1983). Optimal prey selection: why do great tits show partial preferences? *Amin. Behav.* 31:576–584. [8.6, Table 9.1]

Rechten, C., J. R. Krebs, and A. I. Houston. (1981). Great tits and conveyor belts: a correction for non-random prey distribution. *Anim. Behav.* 29:1276–1277. [Table 9.1]

Regelmann, K. (1984). Competitive resource sharing: a simulation model. *Anim. Behav.* 32:226–232. [4.1, 9.3]

Reichman, O. J. (1977). Optimization of diets through food preferences by heteromyid rodents. *Ecology* 58:464–457. [Table 9.1]

Rhoades, D. F. and R. G. Cates. (1976). A general theory of plant anti-herbivore chemistry. In: J. W. Wallace and R. Mansell (eds.), *Interactions between Plants and Insects.* pp. 168—213. Plenum Press, New York. [5.5]

Richards, L. J. (1983). Hunger and optimal diet. *Am. Nat.* 122:326–344. [Table 9.1]

Richter, C. P. (1943). Total self-regulatory functions in animals and human beings. *Harrey Lectures* 38:63–103. [5.5]

Ridley, M. (1984). *The Explanation of Organic Diversity.* Oxford University Press, Oxford. [Preface]

Ringler, N. H. (1979). Selective predation by drift feeding brown trout *Salmo trutta*. *J. Fish. Res. Board Can.* 36:392–403. [Table 9.1]

Robertson. A. I. and J. S. Lucas. (1983). Food choice, feeding rates and the turnover of macrophyte biomass by a surf-zone inhabiting amphipod. *J. Exp. Mar. Biol. Ecol.* 72:99–124. [Table 9.1]

Roitberg, B. D. and R. J. Prokopy. (1982). Influence of intertree distance on foraging behaviour of *Rhagoletis pomonella* in the field. *Ecol. Entomol.* 7:437–442. [Table 9.1]

Roitberg, B. D. and R. J. Prokopy. (1984). Host visitation sequence as a determinant of search persistence in fruit parasitic tephritid flies *Oecologia* 62:7–12 [8.4]

Rosenthal, G. A. and D. H. Janzen. (1979). *Herbivores*. Academic Press, New York. [5.5]

Rothschild, M. and J. E. Stiglitz. (1970). Increasing risk. I: a definition. *J. Econ. Theory* 9:185–202. [6.2]

Royama, T. (1970). Factors governing the hunting behaviour and selection of food by the great tit (*Parus major* L.) *J. Anim. Ecol.* 39:619–668. [Table 9.1]

Rozin, P. (1976). The selection of foods by rats, humans and other animals. In: J. S. Rosenblatt, R. A. Hinde, E. Shaw, and C. G. Beer (eds.), *Advances in the Study of Behavior*, Vol. 6, pp. 21—76. Academic Press, New York. [5.5]

Rozin, P. (1977). The significance of learning mechanisms in food selection: some biology, some psychology, and sociology of science. In: L. M. Barker, M. R. Best, and M. Domjan (eds.), *Learning Mechanisms in Food Selection*, pp. 557–589. Baylor University Press, Waco, Texas. [5.5]

Rubenstein, D. I. (1982). Risk, uncertainty and evolutionary strategies. In: *Current Problems in Sociobiology*, pp. 91–111. Cambridge University Press, Cambridge. [6.4]

Schluter, D. (1981). Does the theory of optimal diets apply in complex environments? *Am. Nat.* 118:139–147. [2.2, 9.1]

Schmid-Hempel, P., A. Kacelnik, and A. I. Houston. (1985). Honeybees maximise efficiency by not filling their crop. *Behav. Ecol. Sociobiol.* 17:61–66. [1.4, 2.6, 9.2, Table 9.1]

Schoener, T. W. (1971). Theory of feeding strategies. *Ann. Rev. Ecol. Syst.* 2: 369–404. [1.2, 1.4, 2.2, 6.6]

Schoener, T. W. (1979). Generality of the size-distance relation in models of optimal feeding. *Am. Nat.* 114:902–914. [3.5, Table 3.1]

Schoener, T. W. (1986). A brief history of optimal foraging theory. In: A. C. Kamil, J. R. Krebs, and H. R. Pulliam (eds.), *Proceedings of the Second International Foraging Conference.* [9.3]

Schultz, J. C. (1983). Habitat selection and foraging tactics of caterpillars in hetero-geneous trees. In: R. F. Denno and M. S. McClure (eds.), *Variable Plants and Herbivores in Natural and Managed Systems,* pp. 61–90. Academic Press, New York. [5.5]

Scriber, J. M. and F. Slansky Jr. (1981). The nutritional ecology of immature insects. *Ann. Rev. Entomol.* 26:183–211. [5.5]

Scriven, M. (1959). Explanation and prediction in evolutionary theory. *Science* 130:477–482. [Preface, 10.3]

Shannon, C. and W. Weaver. (1949). *The Mathematical Theory of Communication.* University of Illinois Press, Urbana, IL. [4.2]

Shettleworth, S. J. (1984). Learning and behavioral ecology. In: J. R. Krebs and N. B. Davies (eds.), *Behavioural Ecology: An Evolutionary Approach,* 2d ed., pp. 170–194. Blackwell Scientific Publications, Oxford. [9.3]

Sibly, R. M. and D. J. McFarland. (1976). On the fitness of behavior sequences. *Am. Nat.* 110:601–617. [7.3]

Sih, A. (1980). Partial consumption of prey. *Am. Nat.* 116:281–290. [2.1, 3.3, Table 9.1]

Simon, H. A. (1956). Rational choice and the structure of the environment. *Psychol. Rev.* 63:129–138. [8.6]

Smith, J. N. M. and H. P. Sweatman. (1974). Food searching behavior of titmice in patchy environments. *Ecology* 55:1216–1232. [Table 9.1]

Snyderman, M. (1983a). Optimal prey selection: partial selection, delay of reinforcement and self control. *Behav. Anal. Letters* 3:131–147. [Table 9.1]

Snyderman, M. (1983b). Optimal prey selection: the effects of food deprivation. *Behav. Anal. Letters* 3:359–369. [6.6, Table 9.1]

Sorensen, A. E. (1984). Nutrition, energy and passage time: experiments with fruit preference in European blackbirds (*Turdus merula*). *J. Anim. Ecol.* 53:545–557. [5.5]

Sparks, J. (1982). *Discovering Animal Behaviour.* BBC, London. [10.4]

Staddon, J. E. R. (1980). Optimality analyses of operant behavior and their relation to optimal foraging. In: J. E. R. Staddon (ed.), *Limits to Action: The Allocation of Individual Behavior;* pp. 101–141. Academic Press, New York. [8.5]

Staddon, J. E. R. (1983). *Adaptive Behavior and Learning.* Cambridge University Press, New York. [1.5, 5.1, 5.3, 6.6, 8.5, 9.5, 10.1]

Stanton, M. L. (1982). Searching in patchy environments: food plant selection by *Colias p. eriphyle* butterflies. *Ecology* 63:839–853. [5.5]

Stein, R. A. (1977). Selective predation, optimal foraging, and the predator-prey interaction between fish and crayfish. *Ecology* 58:1237–1253. [Table 9.1]

Stein, R. A., C. G. Goodman, and E. A. Marshall. (1984). Using time and energetic measures of cost in estimating prey value for fish predators. *Ecology* 65:702–715. [Table 9.1]

Stephens, D. W. (1981). The logic of risk-sensitive foraging preferences. *Anim. Behav.* 29:628–629. [6.4]

Stephens, D. W. (1982). Stochasticity in foraging theory: risk and information. D.Phil. thesis, Oxford University. [4.3, 6.4]

Stephens, D. W. (1985). How important are partial preferences? *Anim. Behav.* 33: 667–669. [2.2, Box, 2.3, 8.6, 9.3]

Stephens, D. W. (1987). On economically tracking a variable environment. *Theor. Pop. Biol.* [4.3]

Stephens, D. W. and E. L. Charnov. (1982). Optimal foraging: some simple stochastic models. *Behav. Ecol. Sociobiol.* 10:251–263. [2.1, 6.4, Box 6.1]

Stephens, D. W., J. F. Lynch, A. E. Sorensen, and C. Gordon. (1986). Preference and profitability: theory and experiment. *Am. Nat.* 127:533–553. [2.4, 3.2, Box 3.1, Table 3.1, 9.2]

Stephens, D. W. and S. R. Paton. (1986). How constant is the constant of risk-aversion? *Anim. Behav.* [6.5]

Stewart-Oaten, A. (1983). Minimax strategies for a predator-prey game. *Theor. Popul. Biol.* 22:410–424. [4.1]

Sutherland, W. J. (1982). Do oystercatchers select the most profitable cockles? *Anim. Behav.* 30:857–861. [Table 9.1]

Takayama, A. (1974). *Mathematical Economics.* Dryden Press, Hinsdale, IL. [7.2]

Templeton, A. R. and L. R. Lawlor. (1981). The fallacy of averages in ecological optimization theory. *Am. Nat.* 117:390–393. [2.1, Box 2.1]

Thompson, D. B. A. (1983). Prey assessment by plovers (Charadridae): net rate of energy intake an vulnerability to kleptoparasites. *Anim. Behav.* 31:1226–1236. [Table 9.1]

Thompson, D. B. A. and C. J. Barnard. (1984). Prey selection by plovers: optimal foraging in mixed-species groups. *Anim. Behav.* 32:534–563. [Table 9.1]

Thompson, J. N. (1982). *Interaction and Coevolution.* Wiley, New York. [5.5]

Tinbergen, J. (1981). Foraging decisions in starlings (*Sturnus vulgaris* L.). *Ardea* 69:1–67. [1.4, 3.5, Table 9.1]

Tinbergen, N., M. Impekoven, and D. Franck. (1967). An experiment on spacing out as defence against predators. *Behaviour* 28:307–321. [Preface]

Townsend, C. R. and A. G. Hildrew. (1980). Foraging in a patchy environment by a predatory net-spinning caddis larva: a test of optimal foraging theory. *Oecologia* 47:219–221. [4.4, 9.4, Table 9.1]

Turelli, M., J. H. Gillespie, and T. W. Schoener. (1982). The fallacy of the fallacy of the averages in ecological optimization theory. *Am. Nat.* 119:879–884. [2.1]

Turner, A. K. (1982). Optimal foraging by the swallow (*Hirundo rustica*, L.): prey size selection. *Anim. Behav.* 30:862–872. [Table 9.1]

Vadas, R. L. (1982). Preferential feeding: an optimization strategy in sea urchins. *Ecol. Monogr.* 47:337–371. [Table 9.1]

Vagner, J. (1974). Optimization techniques. In: Pearson, C. E. (ed.), *Handbook of Applied Mathematics: Selected Results and Methods*, pp. 1104–1180. Van Nostrand Reinhold Co., New York. [10.2]

van Alphen, J. J. M. and F. Gallis. (1983). Patch time allocation and parasitization efficiency of *Asorbara tabida*. *J. Anim. Ecol.* 52:937–952. [Table 9.1]

Vaughan, W., Jr. (1981). Melioration, matching and maximization. *J. Exp. Anal. Behav.* 36:141–149. [8.5]

Vaughan, W., Jr., and R. J. Herrnstein (1986). Stability, melioration and natural selection. In: L. Green and J. Kagel (eds.), *Advances in Behavioral Economics*, Vol. 1. Ablex, New York. [5.3, 8.5, 10.2]

Vaughan, W., Jr., T. A. Kardish, and M. Wilson. (1982). Correlation versus contiguity in choice. *Behav. Anal. Letters.* 2:153–160. [8.5]

Vickery, W. L. (1984). Optimal diet models and rodent food consumption. *Anim. Behav.* 32:340–348. [Table 9.1]

Waage, J. K. (1979). Foraging for patchily distributed hosts by the parasitoid, *Nemeritis canescens. J. Anim. Ecol.* 48:353–371. [8.4, Table 9.1]

Waddington, K. D. (1982). Optimal diet theory: sequential vs. simultaneous encounter models. *Oikos* 39:278–280. [3.2]

Waddington, K. D., T. Allen, and B. Heinrich. (1981). Floral preferences of bumblebees (*Bombus edwardsii*) in relation to intermittent versus continuous rewards. *Anim. Behav.* 29:779–784. [Table 6.1]

Waddington, K. D. and B. Heinrich. (1979). The foraging movements of bumblebees on vertical "inflorescences:" an experimental analysis. *J. Comp. Physiol. A.* 134:113–117. [Table 9.1]

Waddington, K. D. and L. Holden. (1979). Optimal foraging: on flower selection by bees. *Am. Nat.* 114:179–196. [3.2, Table 9.1]

Wallace, J. B. and F. F. Sherberger. (1975). The larval dwelling and feeding structure of *Macroneum transversum. Anim. Behav.* 23:592–596. [1.1]

Wallace, J. B., J. R. Webster, and W. R. Woodall. (1977). The role of filter feeders in flowing water. *Arch. Hydrobiol.* 79:506–532. [1.1]

Ware, D. M. (1975). Growth, metabolism and optimal swimming speed of a pelagic fish. *J. Fish. Res. Board Can.* 32:33–41. [2.6]

Wasserman, E. (In prep.). Psychological aspects of foraging. [Table 9.1]

Wells, H. and P. H. Wells. (1983). Honeybee foraging ecology: optimal diet, minimal uncertainty or individual constancy? *J. Anim. Ecol.* 52:829–836. [9.5, Table 9.1]

Werner, E. E. and J. F. Gilliam. (1984). The ontogenetic niche and species interactions in size-structured populations. *Ann. Rev. Ecol. Syst.* 15:393–425. [7.3]

Werner, E. E. and D. J. Hall. (1974). Optimal foraging and the size selection of prey by the bluegill sunfish (*Lepomis macrochirus*). *Ecology* 55:1042–1052. [2.2, Table 9.1]

Werner, E. E. and G. G. Mittelbach. (1981). Optimal foraging: field tests of diet choice and habitat switching. *Am. Zool.* 21:813–829. [4.1]

Westoby, M. (1974). An analysis of diet selection by large generalist herbivores. *Am. Nat.* 108:290–304. [5.5]

Wetterer, J. K. and C. J. Bishop. (1985). Planktivore prey selection: the reactive field volume model vs. the apparent size model. *Ecology* 66:457–464. [Table 9.1]

Williams, D. (1982). Studies of optimal foraging using operant techniques. Ph.D. diss., University of Liverpool. [Table 9.1]

Williams, G. C. (1966). *Adaptation and Natural Selection.* Princeton University Press, Princeton, NJ. [Preface, 1.1]

Winterhalder, B. (1981). Foraging strategies in the boreal forest: an analysis of Cree hunting and gathering. In: B. Winterhalder and E. Smith (eds.), *Hunter-Gatherer Foraging Strategies*, pp. 66–98. University of Chicago Press, Chicago. [Table 9.1]

Winterhalder, B. (1983). Opportunity-cost foraging models for stationary and mobile predators. *Am. Nat.* 122:73–84. [Box 5.1]

Witham, T. G. (1977). Coevolution of foraging in *Bombus* and nectar dispensing in *Chilopsis:* a last dreg theory. *Science* 197:593–596. [Table 9.1]

Wunderle, J. M. and T. G. O'Brien. (1986). Risk aversion in hand-reared bananaquits. *Behav. Ecol. Sociobiol.* 17:371–380. [Table 6.1]

Ydenberg, R. C. (1982). Territorial vigilance and foraging, a study of tradeoffs. D. Phil. thesis, Oxford University. [7.3, Table 9.1]

Ydenberg, R. C. (1984). Great tits and giving-up times: decision rules for leaving patches. *Behaviour* 90:1–24. [8.5]

Ydenberg, R. C. and A. I. Houston. (1986). Optimal tradeoffs between competing behavioral demands in the great tit. *Anim. Behav.* 34:1041–1050. [7.3]

Ydenberg, R. C. and J. R. Krebs. (In press). The tradeoff between territorial defense and foraging in the great tit (*Parus major*). *Am. Zool.* [7.3]

Zach, R. and J. B. Falls. (1976). Do ovenbirds (Aves: Parulidae) hunt by expectation? *Can J. Zool.* 54:1894–1903. [Table 9.1]

Zach, R. and J. M. N. Smith. (1981). Optimal foraging in wild birds? In: A. C. Kamil and T. D. Sargent (eds.), *Foraging Behavior: Ecological, Ethological and Psychological Approaches*, pp. 95–107. Garland STPM Press, New York. [9.3]

Zahorik, D. M. and K. A. Houpt. (1977). The concept of nutritional wisdom: applicability of laboratory learning models to large herbivores. In: L. M. Barker, M. R. Best, and M. Domjan (eds.), *Learning Mechanisms in Food Selection*, pp. 45–67. Baylor University Press, Waco, TX. [5.5]

Zahorik, D. M. and K. A. Houpt. (1981). Species differences in feeding strategies, food hazards and the ability to learn food aversions. In: A. C. Kamil and T. D. Sargent (eds.), *Foraging Behavior: Ecological, Ethological and Psychological Approaches*, pp. 289–310. Garland STPM Press, New York. [5.5]

Zimmerman, M. (1981). Optimal foraging, plant density and the marginal value theorem. *Oecologia* 49:153. [Table 9.1]

Subject Index

Library of Congress Cataloging-in-Publication Data

Stephens, David W., 1955–
Foraging theory.

(Monographs in behavior and ecology)
Bibliography: p.
Includes index.
1. Animals—Food.   I. Krebs, J. R. (John R.)   II. Title.   III. Series.

QL756.5.S74 1986   591.5′3   86-42845
ISBN 0-691-08441-6 (alk. paper)   ISBN 0-691-08442-4 (pbk.)